The

ART
of
FLAVOR

ALSO BY DANIEL PATTERSON

Coi: Stories and Recipes

ALSO BY MANDY AFTEL

Death of a Rolling Stone: The Brian Jones Story
When Talk Is Not Cheap (with Robin Tolmach Lakoff)
The Story of Your Life: Becoming the Author of Your Experience
Essence and Alchemy: A Natural History of Perfume
Scents and Sensibilities: Creating Solid Perfumes for Well-Being
Fragrant: The Secret Life of Scent

ALSO BY DANIEL PATTERSON AND MANDY AFTEL

Aroma: The Magic of Essential Oils in Food and Fragrance

RIVERHEAD BOOKS

NEW YORK

2017

The

ART

of

FLAVOR

Practices and Principles for
Creating Delicious Food

DANIEL PATTERSON

AND MANDY AFTEL

RIVERHEAD BOOKS
An imprint of Penguin Random House LLC
375 Hudson Street
New York, New York 10014

Library of Congress Cataloging-in-Publication Data

Names: Patterson, Daniel, author. | Aftel, Mandy, author.
Title: The art of flavor : practices and principles for creating delicious food /
Daniel Patterson and Mandy Aftel.
Description: New York : Riverhead Books, [2017] |
Includes bibliographical references. |
Identifiers: LCCN 2017006190 (print) | LCCN 2017016707 (ebook) |
ISBN 9781594634307 (print) | ISBN 9780698197169 (ebook)
Subjects: LCSH: Cooking. | Flavor. | LCGFT: Cookbooks.
Classification: LCC TX714 (ebook) | LCC TX714.P388 2017 (print) |
DDC 641.5—dc23
LC record available at https://lccn.loc.gov/2017006190
p.   cm.

Printed in the United States of America
1 3 5 7 9 10 8 6 4 2

BOOK DESIGN BY AMANDA DEWEY

For my family——D.P.

For Becky——M.A.

Contents

The

ART

of

FLAVOR

Introduction

We met, you might say, at the intersection of scent and taste: a self-taught artisanal perfumer who had been a trailblazer in her work with natural essences, and a self-taught chef known for his creative approach to flavor.

We were introduced by mutual friends in the food business. Mandy came to Elisabeth Daniel, Daniel's first restaurant in San Francisco, bringing her traveling "organ," a fitted case filled with essential oils, which are the concentrated essences of botanical materials, historically used for both flavor and fragrance. We spent the morning smelling them, comparing notes and perspectives. The experience was revelatory. Encountered free of their physical form, the aromas of familiar plants seemed completely new. The essence of fresh ginger was citrusy and piercing. Tarragon revealed layers of complex green and anise notes. Even commonplace black pepper seemed transformed—its essential oil earthy, dusty, and floral instead of merely boldly spicy.

Perfume is, in a sense, disembodied flavor. Abstracted, almost without substance beyond the liquids in which they are distilled, the elements intermingle and create something new. Encountering them in this almost ethereal form makes it easier to pay attention to the relationships among the essences; familiar ingredients reveal themselves in completely new ways. The two of us could have gone on all day, that first day, if we hadn't had dinner to prep and orders to fill.

Not surprisingly, that meeting was the beginning of a long friendship and a fruitful collaboration. As we continued to meet and experiment and work together, we found deep common ground not only in the ingredients we used but also in the ways we created with them: how we selected which ingredients to combine, how we fine-tuned a fragrance or a dish, even how we moved through the process, down to all the little decisions we made along the way. We might each choose, for example, to balance the soft, honeyed sweetness of rose with the bright citrus aspects of fresh ginger. Mandy would start by deciding whether it was going to be a spicy ginger scent with a bit of rounded rose or a buttery floral rose with a smidge of bright spice. Similarly, in the kitchen, Daniel might infuse a custard with the sweet, heavy perfume of dried rose, then lift it with the insistent floral spiciness of fresh ginger. Just as perfume is structured scent, flavor is structured taste.

And the parallels weren't only metaphoric: we both composed with our noses. With no disrespect to Brillat-Savarin, it's actually the nose that serves as laboratory, as much for the chef as for the perfumer. That's why experienced cooks spend as much time smelling as they do tasting. The mingled scent of ingredients describes to the imagination how they might fit together before we actually combine them.

The challenge was in putting words to the process. Good cooks, like good perfumers, learn to orchestrate ingredients into delicious combinations without thinking about it, let alone talking about it. Mandy, however, besides blending fragrances for decades, had also been teaching about perfume composition for years. In the process, she had developed some tools and concepts that helped her students learn how to blend their own beautiful fragrances: ways of grouping ingredients into families that recognize their commonalities while also highlighting their gorgeous singularities, for example, and ways of pairing ingredients so that they "lock" with one another into a whole that transcends the sum of the parts, or so that one ingredient reins in, or "buries," another. Talking with each other, we discovered that these tools and concepts were just as useful in the kitchen. Translating from the language of the kitchen to the language of her studio and back again helped us to make conscious what we think about and do to create flavor.

And that, we realized, is something the many cookbooks we've read and used—even those we've most admired—just don't do. Most cookbooks are collections of recipes, little more. They tell you what to put together, but not why. They are, in effect, the footprints of their authors' process of creating, and there's much to be learned from repeating the recipes in them. But they don't leave you equipped to go your own way.

A recipe doesn't know when you've got sweet, crisp carrots from the farmers' market and a mellow onion, or a few limp tubers and a sprouting bulb from the back of the vegetable bin. A good cook can adapt to either situation and produce something delicious. A recipe can't, for that matter, take *you* into account—your specific desires, tastes, and affinities. It can't factor in your experience, the specific chemistry of your taste buds, your moods and cravings, the bent of your imagination. Recipes can't draw out what you uniquely find delicious—the new thing you want to create out of what's available to you. Great food happens at the intersection of your ingredients and your imagination.

That is why we decided to write *The Art of Flavor*. Cooking is a creative process, and as with other creative pursuits—music, art, writing—there are concepts and tools that can help guide you. Rather than leaving you a set of too-large or too-small or not-in-the-direction-I-wanted-to-go footprints, we wanted to equip you to blaze your own trail. We aim to teach you to become a creative, confident cook who knows how to think about and respond to the ingredients available to you in ways that result in delicious, memorable food.

As two professionals who create sensual experiences for a living, we are used to writing and modifying recipes. As the historian William Eamon writes in *Science and the Secrets of Nature*:

> Recipes collapse lived experience into a series of mechanical acts that, once parsed, anyone can follow. . . . Once it is published, someone else appropriates it, uses it, varies it, then passes it on. At each stop it gains something or loses something, is improved upon or degraded, and is changed to fit new needs and circumstances. Recipes are built upon the belief that somewhere at the beginning of the chain there is someone who does not use them.

This book is designed to make you into someone who confidently adapts recipes to your needs and desires—ultimately, into someone who does not even need a recipe.

So why is it that this book is full of recipes? Because the best way to internalize a principle is to see it in action, and to practice it. The recipes created for this book were designed to demonstrate the principles that govern how flavor is composed. We chose them with a view to showcasing healthy, tasty, inexpensive food made from simple ingredients to which everyone has access. We hope that you enjoy them, but even more, we hope that you learn from them. This book isn't about learning recipes; it's about learning to think about ingredients, process, and flavor in a new way. It's about—as anthropologist Gregory Bateson puts it—"learning to learn." You'll notice that a number of the recipes don't give specific amounts or yields; they're halfway between inspiration and recipe, meant to encourage you on your way, learning to trust your senses and your mounting experience to arrive at the exact amounts that suit your precise ingredients and your precise sense of what is delicious. In the same spirit, each recipe also includes an afterthought—a reflection on how it works, how it might be tweaked to work a bit differently, or another direction you might take it in entirely.

Cooking is about adjustment. It's about paying attention. Cooking is sensual, intuitive, immediate. The most important thing you can do in the kitchen is to trust your senses, which is hard to do when you're trusting a book instead. And once you develop the awareness and confidence to begin to think and cook that way—turning first and foremost to your senses—you can continue on that path. Whether you're tweaking a recipe you find here, in another cookbook, online, in your grandmother's handwritten notes, or creating on your own, you'll be simply following your own instincts and desires as you practice the art of flavor.

A BRIEF, BIASED
HISTORY OF FLAVOR

*Advanced civilizations were possible because there was a
surplus of food, so not everyone had to farm all the time.
Advanced civilizations are where cooking for survival
changes to cuisine—cooking with awareness,
for a purpose other than just to make food edible.*

LINDA CIVITELLO, *CUISINE AND CULTURE*

It's clear that people have been eating well—and cooking well—for centuries. The evidence isn't only in old cookbooks, which are a fabulous paper trail of their own. It's also in novels, art, poetry, and song. But what exactly people think defines good food—and good cooking—isn't easy to tease out, because it's always been bound up in broader cultural notions about what is familiar and what is exotic, what is healthful or harmful, what goes together and what doesn't. And of course, it shifts over time. No cuisine is static. Innovations in technology, transportation, and agriculture, currents of trade and migration and urbanization—all of these factors shift cultural preferences over time.

It was fascinating for us to dive into the history of flavor, and to look for reflections of our own notions about it across time and space. One through-line is the sense that the art of cooking is an expression—and reflection—of a civilization itself. As Brillat-Savarin puts it, with typical pithiness, "Animals feed themselves; men eat; but only wise men know the art of eating." The Dao De Jing asserts that "governing the country is in principle like cooking a small fish," meaning that great care and attention are in both cases essential. Culinary skills were a fine qualification for ministerial appointment. In her essay "The Quest for Perfect Balance: Taste and Gastronomy in Imperial China," historian Joanna Waley-Cohen recounts the legend of the rise of one Yi Yin from cook to trusted minister in China's Shang dynasty—a fast track from gastronomy to governance if ever there was one:

His culinary skills brought him to the king's attention, and in his first audience he transformed the greatest philosophical issues of governance into a menu of foods to be coveted. Among other things, Yi Yin likened the whole world to a kitchen in which one prepares food, and proper government to good cooking. Just as in cooking it was necessary to understand flavours to blend them successfully; so in governing it was necessary to grasp people's sufferings and aspirations in order to satisfy their needs.

The cook has always been part levelheaded administrator. As Michael A. Symons entreats us in *A History of Cooks and Cooking*:

Forget for a moment their mouth-watering creations and think of cooks as rationing resources. Think of them counting out one artichoke for each guest. Think of them balancing the sweet and sour. Think of them ensuring fat, but not too much, and fibre, but not too much. . . . Cooks use their eyes, ears, touch, and, especially, nose, teeth and tongue, to share. And the most balanced results become the most satisfying, those we agree are the most pleasing. We like fairness. Not just through the dishes, cooks conjure harmonious blends out of the social, cultural and physical worlds.

But what, beyond notions of basic evenhandedness, has determined our views on what goes best with what? Of course, before it was possible to ship food easily from one place to another, what was cooked together was largely dictated by what grew together in a given region. In the south of France there was lamb and wild thyme; in Thailand, seafood, lemongrass, and galangal; and in Mexico, corn, beans, and squash. Over time, these combinations became traditional; when people were transplanted to other countries, as travelers and immigrants, they reached for the combinations they'd grown up with, introducing their favorite ingredients even as they adapted to new ones. But cuisines are more than rote combinations. They gradually evolved overarching principles that attempted to impose structure on how to bring ingredients together in harmony.

As in our own time, ideas about what constitutes good food have always been entwined with rules governing the health of the body. For example, Taoist theories dictated that foods should achieve a balance between yin and yang. Yin is cool, dark, moist, and associated with the feminine; hence yin foods—green vegetables and creatures that live in the water—are considered cooling. Yang is hot, bright, dry, and associated with the masculine, and yang foods—fatty and spicy and piping-hot foods, for example—are considered heating.

Similarly, the medieval practice of balancing humors—which had its roots in philosophical and medical concepts from ancient Greece—held that the universe was made up of four elements: fire (hot and dry), water (cold and wet), earth (cold and dry), and air (hot and wet). The human body had four related fluids or humors: choler or yellow bile, phlegm, black bile, and blood. The aim was to eat so as to balance the individual's humors, achieving an optimal state of warm and moist. Disease was understood as the result of humoral imbalance and was to be avoided (or treated) by adjusting one's diet, as described by medieval food scholar Ken Albala in *Eating Right in the Renaissance*: "Cloves would bring into balance the excessively phlegmatic person. . . . Conversely, a sanguine youth should abstain from wine because it would only increase his natural imbalance toward heat and moisture."

As Paul Freedman explains in *Out of the East*, the ideas that animated such principles of "healthful" eating were not like "the American practice of having a diet soft drink to offset a cheeseburger. Rather it is a notion of harmony and complementarity, linking foods and ingredients that belong together for reasons of both taste and balance, or even that medical balance is what lies behind the achievement of beautiful gastronomic effects." In fact, under the trappings of healthfulness, deliciousness often seems to be the real point. Albala underscores this:

> The key to understanding the qualities in the humoral system is flavor. Behind nearly every single qualitative evaluation is ultimately a taste test, and flavor is the most consistent criterion for categorizing foods. . . . Everything can be placed into one of seven basic flavor categories: sweet, bitter, acute, salty, acidic, styptic, and unctuous. Most would add an eighth as well: insipid. The Hellenists also added "acrid" as the hottest of flavors, associated with pepper and mustard.

Mandy had done a deep dive into the history of perfume for her previous books on perfumery, and had been entranced by the sense of stepping back into epochs when the lines between enterprises—cooking, medicine, worship, adornment—were not drawn as they are now. The very notion of cooking as a discrete activity is a modern invention. In the medieval world, cooking, perfumery, and medicine were entwined. Little distinction was made between end uses of ingredients. Edward Schafer makes this point in *The Golden Peaches of Samarkand*:

> Just as no hard and fast line can be drawn between cosmetics and drugs in the civilization of the medieval Far East, so any attempt to discriminate precisely between foods and drugs, or between condiments and perfumes, would lead to frustrated

misrepresentation of the true role of edibles in T'ang culture . . . spices and perfumes had their parts to play in religion as well as in medicine, and also in daily life, to preserve food, to repel unpleasant insects, to purify noxious airs, to clean the body and beautify the skin, to evoke love in an indifferent beloved, to improve one's social status, and in many other ways.

Medieval apothecaries, cooks, and perfumers alike used musk and ambergris and civet and rosewater along with precious spices to bring both exquisite fragrance and extraordinary flavors to their concoctions. Their recipes were jumbled together in volumes that were sometimes called Books of Secrets. Yet there is no doubt that the sensual orchestration of ingredients was a driving force in the composition of such concoctions from their earliest manifestations—and the systems that emerged in many cultures for categorizing them reflected this. In cooking during the Arab Empire, as Bernard Rosenberger notes in the essay "Arab Cuisine and Its Contribution to European Culture," fragrances were "the noblest of all food additives." Those derived from animals headed the list, chiefly musk and ambergris, although their costliness meant that only the very rich could afford them. Also prized, and more readily procured, were rosewater, saffron, cinnamon, galangal, clove, mastic, nutmeg, cardamom, and mace.

We were delighted to discover an unapologetic sensualist in Ghiyath Shashi, the sultan of Mandu (near present-day Mandav, India). An eccentric with some very modern interests and attitudes, he is the original author of *The Sultan's Book of Delights*, a collection of recipes for cooking and perfumes that dates to the late fifteenth century. A lone copy of it survives, in the British Library, but versions of many of the recipes it contains are still in use today. Upon assuming the sultanate in 1469, Ghiyath Shashi immediately announced that, having supported his father, Muhmud Shah, for thirty-four years, he decided not to extend his kingdom or spend his time on the cares of state. He left that to his son, Nasir Shah, while he gave himself up to seeking pleasure, a pursuit that he hoped his subjects would share.

The recipes—to which Nasir Shah later added—reflect a culture that literally feasted on aromatics. Many concoctions are recommended both for perfume and for flavoring—instructions to "eat this or rub it on the body, or put it in food" are not uncommon. Readers are exhorted to "rub rosewater and musk onto their private parts and in their mouths and to put sandalwood on their throats. Essence of musk is good for the mouth, [also] put aloes perfume into the mouth. Rub rosewater on the forehead, sniff flowers, scatter spikenard on the head, rub saffron on the face, use scented flower oils of every kind, make scented powder with the sweet scent of flowers, polish the two front teeth, rub perfume into the handkerchief, wash the whole body with rosewater." Other tips: "Fill pockets with musk and sew them up. Rub scented paste into every belt and into the armpits." For fresh breath, "hold a white China rose in one side of the mouth and turmeric leaves in the other side."

The recipes are a marvel of specificity and creativity, reflecting a deep understanding of ingredients and the nuances that distinguish them from one another, all fueled by a sensual delight in creating delectable concoctions. They are careful to indicate when to use dried ginger and when fresh, or when to roast cumin with salt. One recipe calls for both sweet orange peel and sour orange leaves, and another recommends putting dough that has been fried in ghee "amongst roses so it acquires a sweet smell." Not a page passes without mention of musk or ambergris, and often a recipe specifies white ambergris or black. Methods for stuffing limes and oranges capitalize on the essential oils that are concentrated in the peels of citrus fruits. The attention to the details that make flavor exquisite extends to the serving suggestions, such as the recommendation to present a lime sherbet seasoned with pepper in glasses that have been scented with aloeswood.

It was this kind of sensual, sophisticated engagement with a range of alluring ingredients that made us sit up and take notice. We felt like we were discovering kindred spirits who thought about flavor the way we do. From an earlier period of the history of what is now modern-day India had emerged the concept of *rasa*. Reading about it was like finding a fellow traveler, or a long-lost lover. We quickly recognized that the concept comes close to the way we think and talk about flavor—centuries ago and a world away. The concept refers both to the essence of an ingredient, its purest and finest part, and to the pleasure one takes in experiencing that flavor. Moreover, rasa conceives of both the flavor of a dish and the process that achieves it as irreducible. "Rather than thinking of flavoring as an enhancing additive," writes Susan L. Schwartz in *Rasa: Performing the Divine in India*, Indian traditions view flavor as an essential, defining quality of food. "One does not add herbs and spices as a separate and intriguing supplement but as a part of the process of creation."

Rasa acknowledges the entirely new flavor that can emerge from a combination of ingredients: "The standard analogy is that of a blend of a basic food, such as yoghurt, with a number of spices; the resulting substance has a unique flavor (*rasa*) which is not identical with any of the single elements comprising it," writes Donna M. Wulff in "Religion in a New Mode: The Convergence of the Aesthetic and the Religious in Medieval India." *Rasa* is about creating flavor in a way that is integral to a dish, not an afterthought or adornment—it puts flavor at the center of cooking and eating.

Cultures with this kind of thoughtful and intricate engagement with flavor seem to use the widest array of ingredients, and appear particularly alert to the possibilities presented by the most intensely aromatic of them—the herbs and spices and flowers and citruses. The carefully orchestrated use of spices is a game-changer in the mastery of flavor, and Indian cooking is rightly celebrated for that.

You may be surprised to learn that for a long time even Western cuisine was more adventurous than its reputation would lead you to believe. Herbs, of course, grow locally the world over and so were always at hand for use in food and medicine alike, and some spices, such as anise, fennel, and coriander, were widely available as well. Yet we tend to forget how early the more exotic spices became available in the West, and how popular they were. With their penetrating aromas that could survive long voyages, they mounted a multipronged attack on the senses that few could resist. They could be worn on the body as perfumes, season food, or fragrance your home as incense. Part of their appeal, too, was their exotic origins. They came from fabled, faraway lands and carried the whiff of all that was beautiful and rare. As Paul Freedman notes in *Out of the East*, spices appeared in 90 percent of the recipes in medieval

English cookbooks and in 75 percent of the recipes in European cookbooks from the thirteenth to the fifteenth centuries—a proportion that covers both sweet and savory dishes (a distinction that, for that matter, did not yet cleanly exist).

Sugar was considered a spice in its own right—a way of making medicinal concoctions with bland or unpleasant tastes more palatable. It was also valued for its ability to influence texture, whether as a syrup or boiled to candy. By the middle of the seventeenth century, the increased availability of sugar from the colonies had paved the way for the dichotomy of dishes into separate categories of sweet and savory, and sugar began to be limited to what we now think of as "sweets," with a higher concentration of sugar used in fewer kinds of dishes. Sugar also played a decisive role in the popularity of coffee, chocolate, and tea—inherently bitter substances that were initially introduced into Europe as medicines and didn't become widely popular in the West until they were sweetened with sugar.

So what happened? How did Western cooking—the cooking of the English-speaking world in particular—lose its polymorphous sense of flavor and become known for a missionary-position-like blandness and monotony? There were many factors, and many doctoral dissertations and colloquia have been devoted to them. We are not food historians, but a few developments are worth pointing out. First, the tastes of the northern European nobility—those wealthy enough to afford more than a subsistence diet— began to change, with France leading the way. The appearance of François Pierre, Sieur de La Varenne's seminal *Le cuisinier françois* in 1651 marked a major turning point in the history of flavor in the West. La Varenne was unapologetically uninterested in promoting healthful eating. He championed balance and subtlety in flavor. Culinary trends veered away from spices and toward indigenous foods such as truffles, olives, and herbs. Recipes were based on reduced meat bouillons, eggs, and cream, along with the highly adaptable flour-and-butter roux. Sauces that had been thin and heavily spiced turned thick, elegant, and savory. Vegetables became more popular. Subtlety trumped strong flavors. Slowly this new kind of cooking spread across Europe and, via the restaurant, trickled down to the middle class, and the new Euro-minimalism became codified into books and recipes and menus.

Obviously, the spices that had once been so highly prized by European cooks did not suddenly taste different. But taste is not only about flavor. In fact, the word *taste* is derived from the Latin word *taxare*: to touch, to value, to judge. Its culinary meaning is secondary, overlaid. More broadly, taste came to indicate a preference, and eventually a value judgment. Advances in medicine meant that spices were no longer considered essential to good health. And thanks to the advent of mass cultivation and production, burgeoning sea trade, and constant colonialism (with its attendant atrocities), spices had become inexpensive. What were once luxury items were now available to everyone. As spices lost their status as a symbol of luxury and wealth, the taste of the upper classes for them declined as well.

Some spices—chiefly pepper, saffron, ginger, cloves, and nutmeg—continued to be used, but in smaller amounts and in concert with, or in deference to, herbs. Cinnamon was largely relegated to sweets. "French dictionaries of the period began to distinguish between medicinal spices and aromatic spices, and instead of indicating the dietetic utility of the latter, they warned against immoderate use in cooking," observes the French historian Jean-Louis Flandrin in the essay "From Dietetics to Gastronomy: The Liberation of the Gourmet." By the end of the eighteenth century, spices in European cooking were no longer in good taste.

On the one hand, cuisine was theoretically free of allegiance to any dictates other than what was pleasurable, and anyone could judge it on those grounds, an attitude best expressed by the eighteenth-century statesman the Abbé Dubos, as Carolyn Korsmeyer quotes him in *Making Sense of Taste*:

Does one reason in order to know if a ragout is good or if it is bad, and does it ever occur to anyone, after having posed the geometrical principles of flavor and defined the qualities of each ingredient which makes up the composition of foods, to discuss the proportions of their mixture, in order to decide if the ragout is good? One never does this. We have in us a sense designed in order to know if the chef has followed the rules of his art. One tastes the ragout, and without knowing the rules, one knows if it is good.

On the other hand, what this new democratic-seeming attitude appeared to result in was allegiance to a pretty narrow idea as to what constituted good food. French cooking, and the values it represented, was exported around the world, a culinary colonialism that often disrupted native cuisines. It quickly embedded itself in the United States, and to this day many Westerners' favorite cuisine remains that of the era of classical French cooking, still celebrated in a dwindling number of jacket-and-tie restaurants in major European and American cities.

In the meantime, out of their jackets and ties, Westerners were consuming an ever-growing quantity of processed food, and along with it, an idea of flavor as just another quality that could be injected into ingredients, rather than a way of orchestrating their inherent properties. The introduction of artificial scents and flavors around the beginning of the twentieth century had created an opportunity for a new field of endeavor: the flavor industry. If flavor could be whatever we wanted it to be, what the food industry wanted was for flavor to be profitable.

The flavor industry was not entirely new, of course. In a sense, the earliest instances of deliberately flavoring food with herbs and spices can be considered to be part of it. Flavor as a distinct industry, however, began to emerge with the technologies that allowed the extraction of essential oils from plants. And as with the development of cuisine in general, it was entwined with the development of the fragrance industry. In fact, to this day the huge conglomerates that create fragrances for the perfume industry also create flavors for the food industry.

Essential oils are where the flavors and fragrances live in plants—the volatile aromatic compounds that make up the way we experience them when we smell and taste them. In the case of citrus fruit rinds, the oils can be extracted by simple pressing. Most other plant materials have to be put through a process of distillation with water or steam, being heated in a still to separate the oils from the plant material. There is evidence that the ancient Persians employed distillation as far back as 3000 BC. Around 1000 BC, the Arabs rediscovered the process, and from that point forward it was continually refined. The method depends on the fact that aromatic substances—chemicals—can be removed from plant materials by volatilizing them with steam, and collecting and condensing the vapor. Provided the volatile substances are not soluble in water, on cooling they will separate from the watery distillate and can be removed and preserved in a relatively pure condition. In direct distillation, the plant material comes in contact with the boiling water. Steam distillation is a gentler method of extracting essential oils, and more widely used. In this method, steam is generated in the still, or in a separate boiler, and blown through a pipe in the bottom of the still, where the plant material rests on a stack of trays for quick removal after exhaustion.

The history of using essential oils for flavor is documented in European cookbooks up to three centuries old, in recipes for bitters, cakes, puddings, cordials, sauces, gum, cocktails, colas, ketchup, candies, cookies, and more. (There are also Arab and Mughal examples from many centuries earlier.) At first, the available flavors were relatively few, limited by the state of technology. Up to the sixteenth century, only a small group of essential oils were distilled and in use, among them frankincense, cinnamon, sage, rosemary, rose, and cedarwood. But in the seventeenth century, with advances in the craft of distillation, a plethora of other essential oils joined the mix: cardamom, anise, angelica, lovage, mace, nutmeg, caraway, fennel, pepper, juniper berry, basil, thyme, lemon, coriander, dill, oregano, chamomile, spearmint, cumin, cloves, orange, saffron, and wormwood.

Along the roadways of nineteenth-century England, many establishments were named after the itinerant sellers of essential oils—as attested to by the many "Green Man" tavern signs of the day. Their products were initially based on the natural extracts from herbs, spices, and resins that were already being used in the fragrance and pharmaceutical industries. As methods of extraction were refined to allow the isolation of individual aromatic components, these isolates also became useful flavor materials. Soon the flavor industry overtook the perfume industry as a consumer of essential oils, especially those derived from citruses, spices, and mints.

But in its modern incarnation, flavor was increasingly isolated from and even synthesized without actual food. Beginning with coumarin in 1868, chemists started to figure out how to synthesize some of the specific aroma molecules they had isolated in the distillation process, for use in both flavor and fragrance. Initially these were used as a supplement to natural ingredients, to add aroma and taste to beverages and candies. But they quickly supplanted the use of essential oils for the same reasons that synthetics were taking over the fragrance industry: they were cheaper and more reliable. They were also of lower quality, as they could not replicate the complexity of flavor as it occurs in nature, where the fragrance of any given plant is composed of a complicated bouquet of dominant and trace molecules. By the 1960s, however, the cruder synthetic ingredients were being prominently used to re-create a smorgasbord of flavor offerings, from fruits to vegetables to meats. Today there are more than two thousand synthetic materials in routine use for flavor, and another three thousand in the experimental stage.

For more than a century—even before its turn to synthetics—the flavor industry's ways remained almost entirely cloaked in mystery, its formulas considered trade secrets. It wasn't until the 1960 publication of the industry bible, *Food Flavorings: Composition, Manufacture, and Use*, by Joseph Merory, that the secrets of their formulation began to become widely known. A commercial formula for ketchup was disclosed to be made with the essential oils of clove, cassia, nutmeg, pimento berry, mace, and celery seed; sweet pickle was flavored with cinnamon, cassia, black pepper, coriander, caraway, pimento berry, and clove; ginger ale featured ginger essential oil, of course, along with orange, lime, mace, coriander, and, remarkably, rose, to smooth out the rough edges.

Sodas, not surprisingly, were a major focus of the emerging flavor industry. Soda syrups were originally artisanally concocted, to be served at opulent soda fountains that often featured marble counters adorned with mirrors and trimmed with gold, establishments that rivaled saloons as meeting places. In 1911, according to Darcy O'Neil's popular history *Fix the Pumps*, the United States had more than a hundred thousand soda fountains that served more than eight billion drinks a year; now fewer than a hundred remain. The fountain was presided over by the "soda jerk," whose job it was to squirt syrups into the sodas. Essential oils—ranging from clove, orange, and peppermint to rose, musk, civet, and ambergris—were used in the formulas, many of which seem playful and inventive today. They were often created by pharmacists and considered to be medicinal, which increased the need for secrecy. Coca-Cola, for example, was originally formulated to treat dyspepsia and headache and included lemon, lime, orange, cinnamon, and nutmeg essential oils (not to mention cocaine).

The Food and Drug Administration recognized the maturity of the flavor industry by giving it a kind of benediction, via the Food Additives Amendment of 1958, a list of substances "generally recognized as safe" (commonly referred to by the acronym GRAS). Formulas using only substances on this list were not required to be officially tested before being added to foods and drugs. The GRAS list is constantly updated as testing results become known, and to this day it includes many essential oils as well as synthetic substances that can be used for flavoring.

The flavor industry really took off after World War II, when the interest in artificial flavors, and the money to be made from them, burgeoned along with the rise in processed foods and the profitability and shelf life they promised. It had become clear that the decline in nutritional richness that was a by-product of excessive processing was accompanied by a decline in flavor richness, and most of the flavor industry's mission became how to replace that missing flavor, just as the vitamins depleted by all that food processing were now being added back into cereals. Essential oils continued to play a part in creating flavor, but now they were used the way synthetic oils were. Individual molecules were isolated so that they could be mixed together with other isolates or synthetic oils to create the simulacrum of a naturally occurring flavor, with none of the unruly complexity of the thing itself. Nor were natural essences generally used to contribute the rich flavor of the plants from which they are derived. Basil essential oil could have easily been used much as you would use fresh basil—as a delicious addition to a canned tomato sauce, for example. But it is more likely to be sent down the slippery slope toward Frankenflavors—as part of a host of components to create artificial berry flavor, say.

As with other industries, the flavor industry has consolidated itself into a handful of international conglomerates. In the 1920s there were about seventy family-owned essential oil and aroma chemical-flavor companies in the United States, over fifty of them located in lower Manhattan. By the 1970s more than three-quarters of them were out of business. With the help of gas chromatography (GC) analysis, such companies can identify and reproduce any number of the hundreds of molecules that create flavor and fragrance in a given plant. On the printout of a GC analysis of, say, actual basil oil, a wide range of its component molecules appears. But in practice, only a dominant handful are selected for inclusion to approximate the flavor of the original. By limiting themselves to "only those notes deemed essential to a flavour's characteristic taste and smell, they are able to produce a heightened sensation of that flavor," write Constance Classen, David Howes, and Anthony Synnott in "Artificial Flavours." Artificial flavors as we know them today, they say, "are consequently at once much less than their originals and much more. Our contemporary craving for larger-than-life flavour is reminiscent of the medieval appetite for spices. While spices brought medievals a taste of Eden, however, artificial flavours are reminiscent rather of Disneyland, a synthetic paradise of consumer delights."

Artificial blueberry is a good example of the limitations of fake flavor. Today, blueberries are regarded as a "superfood" packed with antioxidants and flavonoids, high in potassium and vitamin C. They are believed to help lower the risk of heart disease and cancer and to reduce inflammation in the body. None of these remarkable attributes accrue to fake blueberry flavor, but the positive associations with blueberries have driven flavorists to jump on the bandwagon and produce artificial blueberry flavor to further increase the marketability of "healthy" foods such as vitamin water, yogurt, and tea.

Neither, of course, does synthetic blueberry flavor replicate the dimensionality of actual blueberry flavor. The key profile descriptors for actual blueberries, according to flavorist John Wright in *Flavor Creation*, include *fruity, floral, green, pungent, cheesy, vanilla, candy, medicinal, buttery, citrus, cooked, nutty, sharp, herbal, sulfuric,* and *vanilla.* As Wright observes, blueberries have a very complex flavor profile, dominated by fruity, damson plum–like damascenone, apple-like ethyl 2-methyl butyrate, and lavender linalool. "These three notes put together," however, "would only give a sketch of the profile. The true character depends on a complex mix of secondary notes added in relatively small quantities, which are the facets of the ingredient. Preparing a delicious-tasting blueberry flavor is very challenging." But no matter how many facets the laboratory troubles to include, it will never match the depth of flavor of a blueberry itself.

The most recent trend to emerge from the flavor industry is what's known as flavor-pairing theory. It originated from the same gas chromatography techniques that allow chemists to analyze the molecular structure of flavor ingredients with a view to replicating aspects of them. GC made it possible to see what chemical components different foods have in common. This led to the theory that foods that share key flavor components can be substituted for one another, will pair well with the same foods, or, most radically, will harmonize well with foods that share the same major flavor molecules. Some of the combinations that derive from this theory are unusual, to say the least: strawberries and Parmesan; poached banana and ketchup; chocolate and caraway.

The flavor-pairing approach has spread to the world of oenology and molecular gastronomy, as chefs and sommeliers search for new ways to combine ingredients and dishes. But as we shall see, such an approach overlooks the fact that the commonalities such foods have on a molecular level, because of similar dominant aroma molecules, are far eclipsed by the trace elements that distinguish them. A pineapple, in other words, is not necessarily an apt substitute for a strawberry, as a strict flavor-pairer might have you believe. While the two foods share dominant aroma molecules, the caramel-like furaneol and the clovelike eugenol, they also have wildly divergent ancillary molecules that, even in tiny amounts, profoundly influence flavor. Nor do they necessarily dance well with the same partners. And because more than just a molecule or two determine a food's flavor, there is no reason to assume that two foods that share a major molecule will combine well. So flavor pairing can be just another culinary crutch that prevents cooks from developing a truly attentive approach to flavor.

Over the past few decades, the pendulum of public opinion has started to swing away from artificial flavors and back toward real food. Many purveyors, chefs, home cooks, and ordinary eaters awakened to what was being lost in the rush for convenience and profitability and joined the Slow Food movement, or were swept along by it. More people began to cook and eat at home, and to eat more adventurously, at home and away. Fueled in part by this resurgence of interest, food itself has become a bigger industry. Restaurants have become more numerous and more diverse. And food is consumed not only as substance but also as entertainment—chefs have become candidates for celebrity, and a plethora of cable channels, blogs, and other outlets are devoted to the subject.

On this tide of popularity, the recipe industry has flourished, with hundreds of versions of any dish available at the touch of a keyboard. But while many of these recipes describe beautiful and delicious dishes, somehow the art of flavor—the understanding that governs which foods go best together, and how—remains underexplored terrain. That is where we are going in this book, and we want to take you with us.

Two

CONSIDER THE APPLE: INGREDIENTS

The only way one can really know things—that is, from the very inside of one's being—is through a process of self-discovery. To know things you have to grow into them, and let them grow in you, so that they become a part of who you are. The mere provision of information holds no guarantee of knowledge, let alone of understanding. It is, in short, by watching, listening and feeling—by paying attention to what the world has to tell us—that we learn. . . . The geologist studies with rocks as well as professors; he learns from them, and they tell him things. So too the botanist with plants and the ornithologist with birds.

TIM INGOLD, *MAKING: ANTHROPOLOGY, ARCHAEOLOGY, ART AND ARCHITECTURE*

Consider the apple. Actually, consider two apples: the Granny Smith and the Gala. The Granny Smith is green, firm, heavy for its size. The Gala is smaller, red, softer—when you squeeze the skin, the underlying flesh gives easily. Now take an imaginary bite. What is the experience? The Granny Smith is tart and densely crunchy, with a clean, rather nondescript flavor that vaguely suggests green, medicinal herbs. The Gala is sweet, with uplifting vanilla-floral aromatics. The two ingredients don't look the same, feel the same, smell the same, taste the same. There is no "the apple." There's not "the Granny Smith" or "the Gala" either—even two samples of the same variety of apple can vary enormously in the nuances of texture, flavor, and aroma, depending on where they were grown, when they were picked, where and how and for how long they were stored, even which branch of the tree they came from.

We like to think of ingredients as the most basic building blocks in cooking. They are, in the sense that they are irreducible, but they contain multitudes. As the perfumer knows, the flavor of even the simplest ingredient is multidimensional because it's the combined effect of the unique bouquet of volatile compounds the ingredient contains. These vary infinitely in their nature and proportion. As Harold McGee, the unparalleled writer on the science of food, puts it, "With natural materials of almost any kind, both the specific volatiles and the proportions can vary due to the genetics of the plant, growing conditions, harvest time, and post-harvest handling." Any given apple contains more than three hundred separate volatile compounds identifiable through gas chromatography. Even two apples of the same variety, from the same tree, will contain different volatile compounds, in different proportions, if they are at different stages of ripeness. One GC reading of cooked beef revealed 486 such compounds in it, and 541 in black tea. And that fragrant cup of your morning coffee will have in the neighborhood of 800 even before you've added anything to it.

As we'll explore, paying attention not just to the dominant taste of each ingredient but also to its nuances is the key to creating great flavor. In fact, it's telling that we use the same word—*flavor*—to refer to the overall impression of a single ingredient and to that of a finished dish. Both are instances of "out of many, one."

Most people are accustomed to the idea that animal products—meat, fish, and dairy, for example—are dramatically affected by what the animal that produced them ate and drank; indeed, in some gastronomic circles today, this awareness approaches the level of obsession expressed centuries ago in *The Sultan's Book of Delights*, which exhorted, "Buy a yellow cow or a black cow, feed it on sugarcane, green grass, cotton seeds and date sugar and also coconut, nutmeg, cinnamon, pulses, partridge eggs and bamboo leaves." But the taste of plants is equally affected by where and how they are cultivated, not only "heirloom" versus mass-produced tomatoes or peaches, say, but also herbs and spices. These two classes of aromatic ingredients are the most diverse and intense ingredients for creating flavor in cooking, thanks to the essential oils that occur naturally in them—the subject of a later chapter.

Because herbs are leaves, the volatile compounds that distinguish them tend to dissipate quickly as the leaves dry out, which is why the jar of dried basil moldering in your cabinet is a pale shadow of a bunch of fresh basil, let alone the concentrated aromas in a bottle of basil essential oil in Mandy's perfume studio. But spices, too—which retain (and sometimes deepen) their intensity for a much longer time after harvesting—nevertheless vary immensely in character, depending on the most minute differences in growing conditions. You may be aware, for example, that ground cassia bark (*Cinnamomum cassia*), which is the source of most commercially available ground cinnamon, is generally harsher and less dimensional than "true cinnamon" (*Cinnamomum verum*, or *Cinnamomum zeylanicum*). But even the quality of "true" cinnamon is profoundly affected, for example, by where on the tree the bark comes from. The base tends to yield the lowest quality, and the branches that grow in full sunlight tend to yield cinnamon that is spicier than that from branches that grow in shade. The point is not to nose out the exact provenance of every ingredient you use—that's impossible—but to cultivate an awareness of the qualities of the ingredients in front of you, an awareness that should first and foremost drive your decisions about creating flavor.

Just as in perfume, in cooking it's not only the constellation of major aroma molecules that creates an interesting whole; the interaction of even very minor aroma molecules—trace components—can make a significant contribution to the overall character of a dish. They may seem like bit players, but they are crucial to the plot, as we'll see.

The more alive we are to the uniqueness and complexity of ingredients in themselves, the more overwhelming it can be to categorize or even to describe them with any certainty. As Timothy Ingold so aptly observes in *Making*, "Materials are ineffable. They cannot be pinned down in terms of established concepts or categories." And yet if we pay attention, the ingredients themselves tell us what they are: "To describe any material is to pose a riddle, whose answer can be discovered only through observation and engagement with what is there. The riddle gives the material a voice and allows it to tell its own story: it is up to us, then, to listen, and from the clues it offers, to discover what is speaking."

To arrive at this intimate understanding of ingredients, language is key—even when it seems most elusive. "An Yquem paints our palate with frescoes and polyptychs in a hundred gradations," writes Michel Serres of a prized dessert wine in *The Five Senses*. "The eye loses its bearings, as though it were looking into infinity; the mouth tastes until taste itself dissolves; our tongue is lacking in tongues, we do not have fifteen different ways of describing a shade of old rose, our lexicon trembles and stutters, experts invent terms amongst themselves, private and intransmissible." And yet it is precisely the process of evolving a rich, personally meaningful vocabulary of descriptors that lets your imagination penetrate each ingredient, makes you aware not just of its dominant character but of all its nuances as well. Naming something is a step toward becoming more intimate and familiar with it—it's like the difference between saying "that girl" or "my neighbor's cat" and saying "Susanna" or "Stan" (the name of Mandy's beloved late cat)—or better yet, "Susanna Banana" or "Stan the Man."

People who taste for a living, like winemakers and chocolate producers, do avail themselves of a highly nuanced lexicon to describe what they are tasting, and it can seem "private and intransmissible" at times. Yet while wine talk, with its sometimes fantastical descriptions—*notes of aged Cuban tobacco*, or *the aroma of the beach at high tide*—may be a favorite butt of jokes, most of the time it's a good example of the reach and imagination required to really capture what things taste like in all their layers and complexity. In fact, most of us already have some practice making linguistic distinctions to evaluate familiar foods and drinks—especially wine and chocolate!

Perfumers, creating in the rarified realm of pure scent, use language to bring structure and focus to their process. Grappling as Mandy does with natural aromatic materials, in all their idiosyncrasy, makes having a flexible, colorful vocabulary for capturing them indispensable. When she gets a new essence, she will put a drop on a perfume blotter (a small strip of paper), then return to it over a period of hours, noting everything she can smell in it as it first strikes her, evolves, and then fades away. For example, a rose essential oil might start out *jammy* and *rich* and *intensely rosy*, then become *tinged with nutmeg and allspice*, and end on *notes of honey and geranium*.

In reaching for language, professionals of all stripes take our cue from poets, who use metaphor and simile to make feelings and ideas vivid to us. Often, poetic description avails itself of synesthesia, the interconnection of the senses, evoking language associated with, say, sound and touch to describe an aroma, as in these lines from Baudelaire's "Correspondances": *There are perfumes cool as children's flesh / sweet as oboes, green like the prairie.* This kind of cross-fertilization between the senses lends a sensual richness to our apprehension of experience. As we've noted, the word *taste* originally meant "to touch." This makes intuitive sense: tasting involves gathering information about a substance by taking it in, with all your senses.

And of the senses, taste and smell are especially closely entwined—not simply analogous but deeply twined at the physiological root. They are, in a way, two halves of the same sense, which we experience as "flavor" when we eat. Nonvolatile (not aromatic) molecules in the foods we eat are detected by taste buds in the tongue, at the back of the mouth, and on the palate, letting us distinguish among sweet, salty, sour, bitter, and umami, as the dominant tastes are commonly labeled. Recent research reveals that we have additional taste buds for fat; we also sense the heat of a habanero or the minty coolness of menthol in our mouth, though these temperature sensations are more related to our sense of touch (through our trigeminal nerve) than to our sense of taste. All the rest of our experience of flavor—the bulk of it—is our sense of smell, triggered by airborne odor molecules. We detect these odorants at the back of our nose with a small patch of sensor cells, which report their findings to the olfactory bulbs in our brain. The combination of odor and taste information creates our perception of flavor.

Yet for all the rich sensory stimulus food gives us, our language around it has become impoverished. Perhaps as a legacy of an era of processed foods and the synthetic formulas produced by the flavor industry, which injects these foods over and over with the same addiction-triggering attributes, we rarely reach beyond broad categorical descriptors—*sweet*, *acidic*, *salty*, *rich*, *spicy*. Tune in to any of the competitive cooking shows on television and you'll see what we mean: for all our obsession with food, the terminology around it remains incredibly limited and repetitive.

So you should try trailblazing your own terms for taste. It doesn't really matter how subjective your use of them is—in fact, the more subjective the better. Unless you are indeed a professional taster, or aspiring to become one, it's actually not so important that you learn a common vocabulary for talking about food. It's much more important to develop a vocabulary around food that is meaningful to *you*—a vocabulary that helps you become aware of what you taste in a given ingredient, in all its complexity; to imagine it as part of a whole; and then to describe what you taste in the dish as it takes shape and, ultimately, as it is finished. Consistency isn't especially important either; as you get more comfortable around food, you'll find better ways— more alive and specific ways—of describing it.

Ultimately, the narratives you construct around food allow you to remember what you created and to learn from it for the next time. In effect, you are creating a personal database of cooking and eating experiences. Describing what worked and what didn't and what it lacked is the surest way of "learning to learn" in the kitchen.

Following are some ways of thinking and talking more precisely about ingredients that have served us well, to help get you started.

CHARACTER

When you meet someone for the first time, you usually take away an overall impression. It's the same with ingredients—they make a certain general impression, shaped by their most dominant attributes. For example, the character of parsley can be described as *green, fresh, herbal*. Character words tend to be adjectives, and they tend to be broad, capturing major attributes of taste and aroma.

Some character descriptors: *acidic, acrid, aggressive, astringent, baked, bitter, burnt, caramelized, cooling, crisp, delicate, dried out, dry, fatty, fresh, green, harsh, heady, heavy, intense, lackluster, leafy, light, lively, luminous, mellow, musty, piquant, powerful, pungent, putrid, rancid, rich, ripe, roasted, robust, salty, smoky, sour, sparkling, spicy, sweet, tamped down, tarry, tart, toasted, uplifting, vegetal, warm.*

SHAPE

This is another broad way of characterizing the nature of an ingredient that can be helpful when it comes to thinking about flavor. It has nothing to do with physical shape or texture; it has to do with the overall impression of what you are tasting. Perfumers learn to smell in shapes, and cooks learn to taste in them: saffron is flat; cinnamon is pointy. Adding shape to character helps us to think imaginatively—architecturally, to be grand about it—about how it will become part of an ensemble.

Some shape descriptors: *balanced, deep, flat, full-bodied, hollow, pointy, round, sharp, soft, thin.*

{ TEXTURE }

Texture *is* about an ingredient's physical presence—the way it feels in your mouth when you eat it. Texture isn't flavor, obviously, but it has a big impact on how we perceive flavor. The taste of a thin liquid will usually be evanescent, whereas that of a thick, viscous liquid will have staying power. A consommé will feel light and fresh, quickly disappearing on the palate, whereas a puree lingers on the tongue, delivering flavor for a longer period of time. Crunchy granola releases the flavor of the oats and other ingredients in little intense explosions, while creamy oatmeal delivers a comforting warm blanket of melded flavor as well as texture. Putting words to the texture of an ingredient helps you begin to imagine how to cook with it. Fresh, crunchy celery might be best used raw, while old, limp celery would serve better in a soup.

Some texture descriptors: *chewy, coarse, creamy, crunchy, hard, leathery, oily, pulpy, silky, smooth, soft, thick, thin, velvety.*

{ INTENSITY }

Ingredients vary enormously in their flavoring power. Even spices, the highest-intensity category of ingredients, have a range; cinnamon is much more intense than ginger. And intensity varies with the specific ingredient and how it is grown, stored, and processed. Some of the character descriptors listed above (*intense, powerful, delicate*) are about the strength of flavor, but to really home in on intensity it helps to think on a continuum. On a scale of 1 to 10, 1–2 would be what we think of as filler elements: bland flavors like rice and potatoes and some vegetables. The 3–4 range would include most vegetables. Herbs and citrus tend to fall in the 5–7 range. Spices crowd the highest levels of the scale, 7–10. Fermentations like fish sauce and soy sauce and highly spicy ingredients like chilies are at the top of the scale as well.

Intensity is key to thinking about how to structure a dish; it affects the relative proportions of ingredients. In terms of real estate on the plate, intense ingredients are too powerful to be featured as a main ingredient, but they have an outsize impact on flavor, so in that sense they are the stars. Their intensity also merits restraint.

{ FLAVOR FACETS }

As we've mentioned, the flavor of a given ingredient is determined not by one or a few dominant molecules but by an entire constellation of what might be hundreds of molecules, some of them present only as traces. Becoming alert to the unique possibilities of a given ingredient means becoming aware of its nuances as much as its overall character. We call these nuances *facets*. Mandy thinks of them as little wings attached to the ingredients.

As we'll explore in chapter 5, facets are what distinguish individual ingredients among the collections of ingredients with which they are usefully grouped (for example, citrus, herbs). But facets also distinguish one variety of ingredient from a related other, like sweeter, more gentle white grapefruit from sharper, slightly bitter pink, or mellow broccoli from more bitter broccoli rabe, or even two examples of the same variety. Remember, there's no such thing as "the Gala apple."

Naming facets helps you make precise choices when deciding on what kind of dish might use an ingredient to best advantage. All grapefruits have a piney facet, but its intensity varies. Slightly bitter, more intensely piney pink grapefruit might be a stronger candidate for a savory dish where we need more acidity and presence, while white grapefruit might better suit a delicate dessert. Facets also help determine what additions to consider for a dish under construction. Honey or sugar? Both are sweet in character, but refined sugar has hardly any facets, and adds only a one-dimensional sweetness to a dish. That's because the flavor in sugar is in the molasses, which is removed from refined sugar. Adding brown (unrefined) sugar or honey will alter both color and flavor, but brown sugar is warm and earthy, while honey is delicate and floral. Actually, honey itself is wildly variable, depending primarily on the source of the nectar from which it is made; it ranges from mild, flowery clover to deeply perfumed orange blossom to dark, nutty-malty buckwheat.

Appreciating an ingredient in all its individuality—noticing and naming its facets, as best we can—begins to suggest how we might work with it as we set about creating flavor. For example, the sappy sweetness of a fig might remind you of syrup, which could lead you to think that a sauce made with figs would be great on pancakes (it is). Language helps show us where to go.

The words that capture facets tend to be adjectives or nouns that function as metaphors, referencing other kinds of foods or experiences to remind us how they color the dominant character of an ingredient: *airplane glue, amber, anisic, apple, apple peel, baked pears, balsamic, bark, berry, bittersweet chocolate, boozy, bread, brine, brown sugar, bubble gum, burnt wood, buttery, camphorous, candylike, caramel, caramelized fruit, cheesy, citrus, creamy, dark caramel, dirty, earthy, espresso, fat, floral, fruity, grapey, grassy, herbaceous, honey, kippers, leathery, lemon, licorice, liquor, maple, marine, medicinal, milky, minty, moldy, mossy, mushroom, musky, musty attic, nut, oak resin, oaky, ocean, orange, paraffin, peppery, pine, pineapple, plastic, powdery, sap, seashore, seaweed, smoky, steely metallic, strawberry jam, sulfurous, tea, tobacco, toothpaste, vanilla, violet, watermelon, wet dog, wool, zesty peel.*

As you can see, you can get pretty far-out with these! That's fine—this is your own private flavor vocabulary, and the more precise and evocative it is *for you*, the more useful.

How does defining an ingredient—taking in, as precisely as possible, its character, shape, texture, intensity, and facets—suggest where to take it in the kitchen? Consider, say, a butternut squash. Its general character is sweet and vegetal, with a flatness and undertones of earthiness. When cooked, it has a densely creamy texture. These qualities suggest that the squash would be good in a soup. But what else would go well in that soup?

It could use a lift from an ingredient with an angular shape. Fresh ginger fills the bill, and provides an excellent counterpoint. Cooked into the soup, ginger will give the squash dimension and make the flavor warmer and more exciting—but because of its intensity, a little of it goes a long way. Another direction suggested by the flat, earthy, vegetal sweetness of the squash is to combine it with an ingredient that has an uplifting, rounder sweetness, to work with and give dimension to that aspect: for example, a sweet-tart apple with aromatic notes, like a Pippin or a Braeburn, which also brings in a hint of acidity. A vegetable stock would reinforce the vegetal sweetness of the squash and help it hold its own without changing its nature, and a little butter would smooth and unite the elements—as we will see in chapter 8, fat fixes flavor. The citrusy facets of the ginger suggest a direction to go in order to finish the soup memorably: lime zest. Funky, distinctive cilantro would make a great counterpoint to the lime, and a dollop of crème fraîche would pick up on the butter and add a gentle lactic acidity.

The next recipe shows what the process for making that soup might ultimately look like, although of course there are many other directions you might take.

Butternut Soup with Apple and Ginger

SERVES 6-8

1 WHOLE BUTTERNUT SQUASH (ABOUT 2 POUNDS)

1 APPLE, PEELED, QUARTERED, AND THICKLY SLICED
    (PIPPIN, BRAEBURN, OR ANOTHER SWEET VARIETY)

1 QUART VEGETABLE STOCK

4 TABLESPOONS BUTTER

2 TABLESPOONS PEELED, CHOPPED GINGER

FRESHLY GRATED LIME ZEST

2 TABLESPOONS CHOPPED CILANTRO

4 TABLESPOONS CRÈME FRAÎCHE

SALT (UNLESS OTHERWISE NOTED, DANIEL USES FINE SEA SALT)

Roast the squash whole at 350 degrees until it is completely tender; a fork should pierce to the heart of it easily. When it is cool enough to handle, halve it and scoop out the seeds; discard them or save them to roast for other uses. Scoop the flesh out of the skin and put it in a heavy, medium-size pot along with the apple. Add the vegetable stock and butter, and bring it to a boil. Reduce the heat, partially cover the pot, and simmer until the apples are tender, about 20 minutes. Add the ginger and cook for only 5 minutes more—the floral notes of the ginger dissipate if it is cooked for too long. Add salt to taste.

Blend the mixture until smooth with a blender, an immersion blender, or a food processor. Serve hot, garnished with a pinch of lime zest, a scattering of cilantro, and a dollop of crème fraîche.

☞ Consider sprinkling the soup with pomegranate seeds when serving, to provide texture and intermittent bursts of flavor. A smooth soup is homogeneous, so it's a good idea to layer on an element that will break up the monotony.

SHOPPING IS THE FIRST STEP IN COOKING

We hope you're starting to grasp how an intimate, articulated understanding of ingredients helps guide you toward easy, interesting combinations. It also instills a heightened awareness of the ingredients you are buying in the first place, which is where the cooking process actually begins. Even as you put food in your shopping basket, you're making choices that will dictate the possibilities you have to work with and what further choices—of other major and secondary players, of cooking process, of seasoning—will bring out the best in them. Never go by what you think an ingredient *should* taste like, based on its provenance or price tag or other exterior factors. Using all your senses, pay attention to how it *is*.

How do you begin to divine the specific nature of ingredients when you're shopping, since you can't usually taste them? (Though, of course, if they're offering, sample away!) Bring all your sensory apparatus to bear. Are the leaves bright green and crisp or yellowing and beginning to wilt? Is the fish eye clear and firm? Are the tomatoes firm to the touch or yielding?

Above all, smell! Smelling ingredients is both the most accessible way to assess their taste and the easiest place to begin developing a nuanced appreciation of them, because aroma is already so much a part of the way we experience and know about food. We smell when the cookies are done baking, or when the milk has started to turn. Smells tell a story not just about the static state of an ingredient but also about where it's been and what's going on with it. Fresh garlic is sweet and bright; the aroma tells you it would be good used raw. Older garlic tends to smell acrid, moldy, often with subtle notes of rot and decay. This garlic should be cooked, your nose tells you. Same with onions.

Shopping is also a great laboratory for cooks because it allows you to compare ingredients, making distinctions between them rather than considering them in isolation. If you happen on tiny wild strawberries at the farmers' market, for example, smell them alongside the regular kind. Often the tiny wild ones have a distinctive hint of orange in their aroma. Is this cheese too funky? Is that one too tame? Does that chocolate smell waxy in comparison with this deeper, more intense one? Does this melon smell deeply fruity or a little musky? Does that one smell a little pallid?

Try to start with a centerpiece ingredient that you're excited about, a familiar favorite you crave or an unfamiliar one you are dying to try. It is much easier to be inspired—to make imaginative, harmonious choices to create interesting flavor—if you are working with ingredients that you are really interested in eating. But no matter what you begin with—even if it's a chicken approaching its sell-by date or a bunch of green beans that have been sitting in the fridge a day or two longer than you intended—you can make something good with it, as long as you take into account its true nature in its present state.

Consider that many ingredients are really multiple ingredients. Meat consists, broadly speaking, of bones, fat, and muscle—several kinds of muscle. Even the parts that are often discarded, like chicken backs and necks or shank bones, can enrich and flavor stocks and stews. Muscle that didn't work too hard tends to have more fat, less intense flavor, and a delicate texture that benefits from quick cooking. Muscle that has done heavy lifting requires long cooking for tenderness but will yield a more flavorful result—and at a very good price, relatively speaking. That means that you can control how much you spend on a meal simply by knowing how to turn tough, less expensive cuts into a delicious dish. The cooking process for tougher cuts is also far more forgiving, with a much larger window of perfection. For that reason, most of the recipes with meat in this book call for the tougher cuts.

There is also more to plants than the most obvious parts. Don't discard the stems of broccoli in favor of the crowns of florets. If the broccoli is young and tender, the stalks can be thinly sliced and quickly cooked. As the broccoli gets older and the stalks become larger, the skin that protects them will get tougher and thicker, but if you peel it away, the interior will still be tender and bright, and it can be cut and cooked in the same way as the florets. Using more of the vegetable makes your shopping dollars go further, and your cooking tastier.

A bunch of beets in their naturally harvested state includes greens along with the bulbous roots. If you find them in that form, don't discard the greens, which are similar to Swiss chard in texture and flavor. Wash and chop them, then simmer them in just a little bit of water until they are tender. Drain and toss them with the cooked beets and a little fruity olive oil, then add a squeeze of lemon juice to brighten the earthiness of the beets mixture and a few grinds of black pepper to offset their sweetness. Tough leek greens, squash seeds, Parmesan rinds, and citrus peels are other ingredients that can add flavor to your cooking but are too often thrown away.

The beets themselves are a good example of an ingredient whose quality you can't really assess until it is cooked. Try scrubbing the beets and roasting with salt, olive oil, and a little water in a covered pan in a 350-degree oven until tender—about 30 to 40 minutes. When they are cool, peel them and cut them in chunks. Try a bite. Are they densely sweet, or do they taste a little bitter and need a pinch of sugar to bring up their sweetness? More salt? Season them and let them absorb the seasoning as they sit.

Usually you can even work with an ingredient that turns out not to be in an ideal state. Flavor is malleable, to a large degree, and even when perfection is not within reach, deliciousness can be attained through the cooking process and an artful combination of ingredients. But when you find an ingredient that's special—that has flavor to spare—you won't need to work so hard to make it delicious. Moreover, if you pay attention to the qualities that make it special, they will begin to suggest where to go with it.

In an ordinary supermarket, you might encounter a few varieties of cucumber—standard cucumbers of medium size with lots of seeds, giant English hothouse cukes with no seeds, bumpy little kirbies (or pickling cukes) with tough skins, or small Persian cukes with thin skins and minimal seeds. All these suggest different treatments, like kirbies for pickling and English for salads. But in farmers' markets and high-end groceries, variety is making a comeback, and you may encounter "heirloom" and other versions of ingredients you haven't seen before. These present the cook with unique opportunities to notice, appreciate, and make the most of a particular ingredient's signature qualities.

Grocery store carrots, for example, are usually Imperators—the skinny, straight kind that come in one- or two-pound bags. But a blunt-tipped Nantes carrot tends to have a brighter flavor and a snappier texture. When harvested in the cold months and shaved raw, it makes for a great salad. The round, fat, triangle-shaped Chantenay carrot, on the other hand, is usually thick and deeply flavored, and has a meaty texture when cooked; it lends itself to being the star of a main course, perhaps with a bit of meat stock and grains. The range of plant varieties within the same species can be dramatic. Sweet oranges, a designation that includes navel oranges, are in fact usually sweetly indistinct, reminiscent mostly of the taste of morning orange juice. Blood oranges, on the other hand, are sharp and bright, with notes of raspberry. Satsumas are more floral and complex, clementines more densely sweet. Paying attention to the specific qualities of each ingredient will help you choose just the right one for the effect you want.

Given a choice, try to gravitate toward products that seem more alive: brighter, firmer, more aromatic. Time dictates huge variations in the nature of ingredients. With fruit, most obviously, time corresponds to ripeness, which has to do with both sweetness and the complexity of the aromatics. Flavor and aroma generally develop more fully as a fruit ripens. Commercially grown fruit is often picked at a very early stage in order to allow it to be shipped with less damage and to give it the longest possible shelf life. With some exceptions (such as pears and avocados), however, this early harvesting alters the course of the ingredient's flavor development even as it continues to ripen. Tomatoes that are allowed to ripen on the vine before picking simply taste more like tomatoes.

Produce generally degrades in quality with each moment that it's disconnected from the earth. The intensity of its flavor dissipates; texture turns from tender and crisp to mushy or dry or coarse. There's an overall loss of what might be called the plant's life force. For this reason, few herbs dry well; as they dry, they lose the terpene hydrocarbons that give fresh herbs their bright and clean notes. They grow dull and acquire musty and other off notes.

A few ingredients, however, last well or even improve with age. Some older varieties of apples, like Baldwin and Pippin, were bred to withstand harsh winters spent in a cellar, and they actually improve with a few months of down time. Winter squashes tend to need rest time once they are picked, to allow flavor and sweetness to develop. Pears ripen only off the tree.

With many kinds of meat and poultry, aging fosters a complicated enzymatic interaction that breaks down large cell molecules, rendering the flesh more tender and flavorful and often adding cheeselike aromatics. Some meats and fish (gravlax, prosciutto, and speck, for example) are cured by covering them in salt and letting them sit for a time, a process that both preserves them and creates intense flavors. Pork is the most commonly cured meat—salted and aged as ham, smoke-cured as bacon, fermented and cured as salami. Smoking not only helps preserve food but also adds aromatics. Some foods—beef jerky, salt cod—are salted and dried. Asian cooks have long made use of dried scallops and abalone, which develop strong flavors and are used sparingly. Japanese cooking makes copious use of bonito flakes, which are made from skipjack tuna that is cured, smoked, fermented, and dried for many months or years. Bonito flakes have a profoundly meaty flavor, deep and smoky with subtle fish flavors, at once reminiscent of ocean and land. French cooks have for centuries cooked and stored duck in its own fat, as confit. The process preserves it for many months and in addition melds the ingredients into a deep, rich, mellow flavor with a tinge of gaminess. (To a less dramatic degree, the same kind of melding happens when a long-cooked stew is allowed to sit for a day or two.)

In any case, life force isn't just a matter of time. Some plants—like some people!—begin with greater life force because of the kind of soil they're rooted in, the amount of water they get and the nutrients it contains, or simply the variety they are. Countless factors dictate how flavorful the plant will be right out of the ground, and what kind of staying power it will have. Some week-old asparagus may have greater intensity and character than a two-day-old bunch that was cultivated with attention only to ease of harvesting, shipping, and storage.

Of course, some ingredients are meant to last for weeks or even months or years. These ingredients, which constitute your pantry, can be kept on hand, ready to be drafted into service.

Foods like miso, soy sauce, and even hot sauce are created through extended fermentation and are already complicated, composed flavors. During the fermentation process, bacteria and yeasts break down the proteins, carbohydrates, and oils. This process creates the same kinds of chemical reactions that happen when meats are browned, which is why these ingredients can have a meaty quality. Fermentation, in other words, is a cooking process, so these ingredients have in a sense already been "cooked" and bring far more depth of flavor than raw ingredients do. When used to season raw ingredients, they quickly and easily give a dish depth. Add miso to hot water and you have a delicious soup, because so much work has already gone into packing it with flavor. The same is true of cheese and wine and many of the other delicious products of fermentation, which are a great asset in cooking.

These observations about shopping and stocking your kitchen are not meant to be exhaustive. They are meant to make you as aware of ingredients when you are shopping as when you are cooking. Because you already *are* cooking: the process begins with what you choose to begin with. Everything follows from there.

Try this: Every time you go food shopping, practice comparing the aroma of ingredients that are similar, paying attention to the ways in which they are aromatically different—coffee beans from different countries; oolong, green, or black teas; milk or dark chocolate; blood or sweet orange; Pink Lady or Fuji apples. You will begin to notice the different facets of their aromas. Let yourself gravitate toward what draws you, with its vibrant color, its pungent smell, or simply its strangeness. Pick it up and inhale its aroma deeply. Try to put words to its character, the shape of its aroma, its facets. What familiar associations does it call up? A forest floor? What inspirations does it call forth? Maybe think of pairing it with a piney herb? Identifying those bridges from the familiar to the unfamiliar is the key to expanding not just your pantry but also your repertoire. Curiosity may have killed the cat, but it makes for a good cook.

Medieval shoppers had a great incentive to shop with their noses, as they thought that diseases were transmitted by bad smells. They sniffed out John Gylessone, who tried to sell at the market "a putrid sow he'd found dead in a London gutter. . . . [Furthermore] the itinerant pedlar might still be able to pass off his wooden nutmeg and juniper-berry pepper corns on unsuspecting housewives in the countryside" (Reay Tannahill, *Food in History*).

Three

CREATING FLAVOR

The cook makes refinements based on the vicinity of ingredients to one another. Knows how to dissolve liquids into fluids, or solids, poorly cohesive as flesh, into thin or thick sauces, thereby obtaining subtle liaisons. Where does meat end and stew begin? Sometimes even our sense of taste cannot distinguish.

MICHEL SERRES, *THE FIVE SENSES*

What is flavor? Here's the best definition we can think of: a constructed orchestration of ingredients that becomes greater than the sum of its parts. When we say, "What a flavor!" we mean that we are tasting not just individual ingredients but some transcendent synchrony of ingredients that the tongue can't easily reduce to its individual components. Think of that memorable *boeuf bourguignon* you had in the French countryside, a true melting pot of succulent meat, aromatic herbs and vegetables, and mellow wine. Or the mélange of ripe avocado, sweet onion, lime, and pungent cilantro that lock seamlessly together to make a perfect guacamole. Or even the lemon and whiskey that make a good sour. These are all created flavors.

You create flavor yourself when you make your morning coffee. You've just done it so many times you don't think about it.

For starters, there's the bean itself. You no doubt have your favorite: choosing it is the first step in "cooking" it. You may even know that coffee beans are the pit of the berries, most commonly, of one of two types of bush: *Coffea arabica* or *Coffea robusta*. Do you like yours lightly roasted with more floral notes, resulting in a lighter, more acidic coffee? Or do you like them heavily roasted, on the verge of bitter?

You continue the cooking process when you decide how you want yours prepared, even if you've only staggered to your corner coffee shop. Thick, syrupy espresso or lighter Americano or French press? Hot and aromatic or cold and more viscous? With cream that makes it thick and smooth or steamed milk that makes it frothy and dilutes the bitterness with air? Do you take it black, relishing that bitterness; sweetened with sugar or agave; spiked with cinnamon or chocolate or a shot of booze?

Each of these decisions aims to approximate your own idea of delicious. Getting there doesn't require you to know everything about coffee; you just have to know how *you* like your coffee and let imagination and experience guide you. You assess intensity and texture, adding water if necessary. You add cream or milk, sugar or spices in proportions that you know, from experience, to be close to perfect. You stir, smell, and taste, assessing the balance and layering of the mixture in relation to what you've imagined.

Up in the perfume studio, Mandy goes through her own blending process. Here all is calm, ordered, deliciously fragrant. The bottles of essences used to make perfume sit on narrow shelves—the perfumer's "organ"—are grouped together as top notes, middle notes, or base notes. Top notes, which reach your sense of smell quickly, are the most fleeting, and they include orange and lemon, spearmint and tarragon. Middle notes, which could be flowers like rose and jasmine or spices like cinnamon and nutmeg, are used to form the heart of a perfume. Base notes, which are deep and grounding and can linger for days, come from trees and roots and resins, like frankincense and fir. Many labels bear the names of ingredients that are familiar to us from the kitchen—and more than a few of them are, in fact, edible.

A BRIEF HISTORY OF HOT BEVERAGES

Tea

Tea is that singular journey or long labor that, in order to release all its aromas, starts with the selection of the rootstock and lasts until the moment of final preparation, covering a patient wait of at least three years for the plant to produce buds or leaves for picking, and a treatment that brings into play manipulation, heating, oxidation, and a more or less long conservation period. A vegetable growth, it will not release its flavor until it has been first picked, then fermented, and finally infused: these are the three great stages of the synthesis of its aromas.

—DOMINIQUE T. PASQUALINI, *The Time of Tea*

Tea is the perfect illustration of the complexity and variety of flavors found within a single ingredient: more than two thousand varieties come from one plant, each variety determined by how and where the leaves are grown, harvested, and processed. They can be steamed, rolled, fermented, bruised and twisted, and formed into cakes. They can be fresh or aged. The flavor can range from delicate and floral to deep, earthy, and tannic. Legend has it that the origin of the famous lapsang souchong tea was the result of an accident by a planter who needed to speed up the drying process and smoked the leaves over a fire made of pine branches.

continued

Tea began as a medicine and eventually evolved into a beverage drunk for pleasure and sociability. Its aroma is made up of hundreds of molecules. There are many kinds of blended teas, almost all of them flavored, from the orange-cinnamon of Constant Comment to the bergamot notes of Earl Grey. Sometimes tea is scented with flowers, such as jasmine. Sometimes the flavoring is made from essential oils, and other times it has little to do with nature. Classic blends are made by tea blenders who mix known vintages of tea to obtain marriages of flavor, aroma, and color in a process akin to perfumery.

Coffee

Coffee originated in Ethiopia as a food rather than a beverage. The beans, which start out green and have no aroma, were chewed whole or crushed with fat and made into a paste. Over time, people began to infuse coffee in water as a kind of tea. It wasn't until the thirteenth century that they began cleaning and roasting the beans before brewing, and coffee as we know it was invented. Certainly the drink was already being consumed in the Ottoman Empire when Columbus stumbled upon the Caribbean islands, and by the end of the sixteenth century it had been introduced to Venice. By the mid–seventeenth century it was available in English, French, and Dutch cities as well. The widespread availability of sugar from the eighteenth century onward marked a turning point in the popularity of the beverage; so did the establishment of coffeehouses, which provided a convivial atmosphere for socializing. Like tea and hot chocolate, it required copious amounts of paraphernalia to accompany the ritual, which made it a social lubricant and fostered community—as did the intense aroma of the coffee, released when it was roasted.

Chocolate

The ancient Aztecs rubbed chocolate on newborn babies to bless them, and also used it to anoint victims destined for ritual sacrifice. They drank it cold, sometimes adding honey, vanilla, and marigold petals. Chocolate had found its way to Spain by 1544, and a transatlantic trade in the commodity was under way by 1585, when the first shipment of cocoa beans was dispatched from Veracruz to the port of Seville. The Spanish court drank cacao after meals, as we do tea and coffee, flavoring it with sugar, cinnamon, vanilla, and citrus. The ritual of drinking chocolate involved frothing pots, ornate cups, chocolate pots, and trays.

Cacao pods are up to a foot long, with a reddish-brown shell filled with a whitish pulp that contains the seeds. The beans are removed from the pods before fermentation, which takes less than a week. Yeast turns the beans' natural sugars to alcohol and removes their astringent tannins. The beans are then dried before they are roasted, to develop their aroma and flavor and diminish their acidity. The beans are ground to a fine powder and mixed with sugar and either hot milk or hot water, depending on cultural preference.

Away from the heat and noise of the kitchen, where there are so many things to pay attention to—how high should the heat be? is the sauce thick enough? is it time to add the dill?—it's clarifying to consider the alchemy of composition in a distilled way.

Mandy says:

Today I want to make something focusing on jasmine. Jasmine is narcotically floral, complex, heady, and slightly animalic, with a touch of overripe fruit about it.

Creating fragrance is a series of decisions, each of which further limits the choices yet to come. With only the first ingredient on my plate, so to speak, the possibilities for a second ingredient are fairly open: I'm looking for one that will spark a great conversation, much the same way I'd think about seating two guests together at dinner or, even more apt, setting them up on a blind date.

I think about choosing something from the same fragrance family, like ylang-ylang, a floral with fruity banana notes, but I decide that could be boring. The two essences have so many aromatic facets in common they would probably merge seamlessly into each other, locking in a way that cancels out their individuality. The same is sometimes true of ingredients that are not even from the same fragrance family but are similar in another way. Saffron and chocolate, for example, do not smell the same at all, yet they have a similar flatness that tends to result in a lack of buoyancy when they are combined.

On the other hand, I could think about choosing an essence that offers a strong contrast to the jasmine, to introduce a counterpoint—a frisson of excitement—from the get-go. These kinds of combinations can be thrilling when they work; they are also capable of creating the biggest disasters: muddy, strange, unbalanced. Starting with strongly contrasting ingredients makes it important to keep close control over the remaining ingredients, so that they build a bridge between the two disparate elements.

I decide to go for the contrast and choose grapefruit as my supporting character. I smell the jasmine and the grapefruit together, imagining how the sparkle of the grapefruit could brighten the sultry jasmine. Now every other choice I make must capitalize on the dynamic of this duo.

Like eating a beautifully constructed dish, smelling a perfume is a dynamic experience, not one uniform thing. First the top notes lift off, followed by the middle notes and finally the base notes. I have to be thinking about that evolution as I blend. A well-constructed perfume is an orchestrated whole, into which each piece needs to fit seamlessly. I have started building this perfume with a middle note of jasmine and added a top note of grapefruit. Now for the base note I choose oakmoss, a lichen that grows on oak trees. It is rich and loamy, and its "foresty" facet will help ground the perfume, balancing both the sweetness of the jasmine and the brightness of the grapefruit.

What else do I have to think about? Proportion: I will need to use more of the grapefruit than the intense jasmine if I want the grapefruit to be able to hold its own. Or maybe I'll allow the grapefruit to be "buried"—to let the jasmine mute or eclipse it, so that the combination of the two is surprisingly less than would be expected. I sometimes deliberately let a spicy note be buried by a floral note, say nutmeg by rose, so that the edge of the rose seems only to be feathered with a bit of spice in the finished blend. All these effects have their parallels in flavor.

Now, how to finish the perfume? When balancing essences with strong personalities, I sometimes use "filler notes"—essences with low odor intensity that don't bring much of their own to the perfume but play well with others and help bind more assertive ingredients together harmoniously. They are the negative space in a painting, the "easy" guests you can seat with anyone. At the opposite end of the spectrum are the "accessory" notes defined by their high odor intensity. Even a smidge can dominate a perfume, but when they are sensitively dosed, they bring nuance and complexity to a fragrance, nothing short of magic. Cinnamon, coffee, and clove are all accessory notes, much as they are in cooking.

I work drop by drop, adjusting based on how my nose tells me the perfume is unfolding, the same way you'd sniff and taste the tomato sauce to decide whether you want it to lean toward basil or rosemary, and how far, or whether it needs a pinch of sugar to counter its acidity.

Now Daniel's turn:

A dish, like a perfume, begins with an ingredient—an ingredient that must be balanced with other ingredients. But unlike in perfumery—with its clearly defined base, middle, and top notes—in cooking these categories are relative. Herbs and spices generally function as top notes, striking the palate first and defining your initial impression of a dish. Spices usually sit in the middle. But other ingredients play different roles, depending on how they are treated and what else they are combined with. Poached chicken is light, more toward the middle register, but roasted or grilled chicken takes on a deeper and fuller flavor. Roasted vegetables can play base, too, especially in a meatless dish. Most vegetable salads take place entirely in a middle to high register.

Think of a chord. Wherever it falls on the keyboard, it can form a pleasing whole. A bass note gives it depth and persistence. A top note lifts the sound and gives it lightness and delicacy. Middle notes connect top and bottom and add richness. The orchestration is all.

A dish may be inspired by any kind of ingredient—a fresh fish your sister-in-law caught, ripe tomatoes in the garden, a special smoked paprika a friend brought back from a trip to Spain—but as a practical matter it's almost always easiest to start building a dish at the base or middle: with a protein, starch, or vegetable.

Let's say I'm starting with something simple and familiar, like carrots. The first thing to do would be to consider the qualities of the carrots—not just the family traits, but the specific traits these actual carrots possess. Are they fresh and crisp, or have they gone a little limp? Assessing them will tell me if I want to eat them raw or cook them. Are they sweet or earthy? Is their flavor bright or dull? These things will suggest how to season and finish the dish.

The second ingredient in a dish could be anything, but it's wise to let the choice be guided by the carrots themselves. I could put the carrots to roast with other root vegetables—turnips, parsnips, and onions, say—or puree them with sweet potatoes, the flavors melding into something of a mélange.

But what if it's contrast I'm after with my carrots, rather than a seamless blend of similar ingredients? Then I'm thinking citrus.

Carrots love citrus, as it happens—in a universe of very few constant rules, this is one. The exact choice of citrus, however, makes an enormous difference, as does the preparation: Cold or hot, sauté or bake, crisp or tender? Carrots shaved raw in a salad suggest a light hand with seasoning, but roasting—with the deep and rich flavors it develops—invites a more aggressive approach.

If they are mature storage carrots, large and sweet but not that fresh, then perhaps lime would be a good choice to cut through their sweetness and brighten their flavor, especially by including the zest. Carrots tossed with olive oil and salt and roasted until tender, then tossed again with lime zest and black pepper before serving, make an easy and delicious side dish with a fresh green top note.

Orange is another natural pairing for carrots. Orange and orange, sweet and sweet. Carrot juice and orange juice combined, half and half, is a perfect ratio for a drink. A little bit of orange juice added to hot carrot soup just before serving gives it sweetness and more complex aromatics. Added to a chilled carrot soup, in a higher proportion, the orange juice lends a fresh, bright note, perfect for a warm-weather appetizer.

Carrot and lemon is perhaps the most versatile combination. Lemon has some of the sweetness of orange, some of the acidity of lime, and the lowest intensity of all the citruses, which makes it the most versatile of them. If the carrots are used raw, in a salad, lemon coats them as a dressing. When added to carrots that are cooked until tender, it penetrates them fully, giving them an intense sweet-sour flavor.

I consider bringing in a strong accent note. The earthy aroma of ground cumin gives carrots extra depth as they cook, a savory quality. But a little cumin goes a long way; too much and it will overwhelm the flavor of the carrots.

Sometimes combined flavors lock together to create something new. Roasting the carrots on a bed of coffee beans allows complementary facets of the two flavors to become completely married, so that the taste is neither completely one nor the other.

And as with the "filler" notes in perfume, sometimes a finished dish will require stretching the ingredients with neutral elements, which can balance the intensity of the main ingredients without altering them significantly. Potatoes or rice in a stew function that way. So does vegetable or chicken stock—with which roasted carrots might be turned into a soup.

All these decisions have a bearing on the notes you will choose to finish the dish. Maybe the soup wants some yogurt on top for tang, fruity olive oil and herbs for freshness. A squeeze of juice to complement the zest on the roasted carrots, a pinch of salt.

Carrot and Sweet Potato Puree
SERVES 4

This sweet, rich vegetable side makes an interesting, delicious alternative for Thanksgiving. Carrots and sweet potatoes have considerable overlap in character—sweet and round—but with slightly different facets: the carrot a little fresher, and the sweet potato a little earthier. Such similar ingredients typically should be combined in relatively even proportions. If you use mostly carrot, you won't taste the sweet potato; use mostly sweet potato and you won't taste the carrot. In either case, all you'll have done is muddied the flavor of the dominant ingredient. If you want purity of flavor, use just one ingredient. Here, with such similar ingredients, a fifty-fifty ratio—with butter as a medium to help the ingredients connect—creates what you are after: a new flavor that is neither carrot nor sweet potato.

1 POUND CARROTS
1 POUND SWEET POTATOES
4 TABLESPOONS BUTTER
SALT

Peel and slice the carrots and sweet potatoes into even pieces. Simmer them in salted water until they are tender. Drain, then let them sit in the colander for several minutes, until the excess moisture steams off. Rice the mixture back into the pot or mash it, then add the butter and stir until the butter is melted. Season to taste with salt.

☞ A recipe as stripped-down as this one demands extremely flavorful ingredients, lest you end up with a boring dish. If that happens, there are a few

fixes. If the vegetables are not sweet enough, a little bit of sugar will bring out their flavor. A bit of chili heat will lift the combination, as will lemon or lime zest. More on this in chapter 8.

Carrots Roasted with Coffee Beans

It's hard to imagine two more unlikely-seeming plate-fellows than carrots and coffee. One is a denizen of salad bars and stews, the other a morning beverage that on busy days often substitutes for a meal. But the sweet-earthy facets of carrots fit perfectly with the earthy-bitter coffee notes. Carrots have a quiet, grounding aroma, while coffee is uplifting and sharp. Cooked together, the two flavors merge into one new flavor. Dark-roasted beans are particularly good for this treatment, but it is essential that the beans be freshly roasted and of good quality, as the carrots will take on their flavor.

WHOLE YOUNG, FRESH CARROTS

VEGETABLE OIL

SALT

DARK-ROASTED COFFEE BEANS

Preheat the oven to 325 degrees.

Scrub the carrots but keep the skins on. Toss them with just enough oil to coat and sprinkle them with a pinch or two of salt. Place a layer of coffee beans in a heavy, ovenproof pan with a tight-fitting cover, just large enough to hold the carrots, and arrange the carrots on top of the coffee beans.

Cover the pan and roast, shaking the pan occasionally, until the carrots are fully tender, about 45 to 60 minutes. Roasting them fully brings out all their sweetness. Discard the coffee beans and let the carrots cool to room temperature, then slice as desired and serve.

☞ Because of their sweetness and the coffee flavor, these carrots sit between sweet and savory. Combined with yogurt, granola, and a drizzle of honey, they make a fun and sophisticated dessert.

Carrot-Orange Salad
SERVES 4

This salad is a great dish for early winter, when carrots and oranges are at their best, and when vibrant new olive oil is in the stores. It combines cooked and raw carrots, served chilled to accentuate their sweetness. Cold foods register as sweeter than warm foods, but not because the proportion of sugar is actually higher; it's because the perception of salt diminishes more quickly than the perception of sweetness as the temperature of the ingredients drops. Young, fresh carrots work best for this dish. Orange segments add a complementary sweet element and bring out the orangey facets in the carrots, and a vinaigrette spiked with tarragon balances the dish. Shaved fennel echoes the licorice notes of the tarragon, and endive provides bitterness and watery crunch. Green olive oil provides grassy, fresh, peppery tones. This is a simple combination that, through carefully chosen seasonings, turns into a complex blend of sweet, sour, salty, and bitter.

1 POUND CARROTS, PEELED AND CUT INTO ½-INCH PIECES

2–3 SWEET ORANGES

5 TABLESPOONS GREEN OLIVE OIL

SALT

½ POUND CARROTS, PEELED AND SHAVED THIN

1 BULB FENNEL, SHAVED THIN

2–3 HEADS ENDIVE, SLICED INTO ¼-INCH PIECES

2 TABLESPOONS CHOPPED FRESH TARRAGON

2 TABLESPOONS CHAMPAGNE VINEGAR

FRESHLY GROUND BLACK PEPPER

Cover the cut carrots in salted water and simmer until they are tender. While they are cooking, peel the oranges and separate them into segments, working over a bowl to catch the juice. Squeeze the juice from the remaining membranes into the bowl with the segments. Discard the membranes and pith. Drain the carrots, let them cool to room temperature, toss them with the orange juice from the segments and 2 tablespoons of the olive oil and salt to taste, and set them aside. The goal is to infuse the carrots with orange flavor, give them a savory tone, and sharpen the flavors with olive oil and salt. You can do this much a few hours ahead.

To finish the salad, divide the marinated carrots among four bowls and put the orange segments around the carrots. In a mixing bowl, combine the shaved carrots, shaved fennel, and endive with the tarragon, then dress with the vinegar, the remaining olive oil, salt, and black pepper to taste. There should be a healthy tension between sweet and sour, with the salt resting just under the surface. Put the salad on top of the carrots and orange segments.

☞ This salad is fresh and bright, but for more dimension consider replacing the olive oil with almond or hazelnut oil. The nutty, buttery quality of those oils will soften the flavor, making it less peppery and more luxurious. Nut oils are a great way to add underlying complexity, and you can amplify their effect with toasted and chopped nuts.

Roasted Carrots with Curry Powder and Lime Zest

Roasting vegetables concentrates and amplifies their sweetness. The curry powder, added both before and after cooking, bring spiciness and dimension and connects with the bright green lime notes.

6 LARGE CARROTS, PEELED AND CUT INTO 1-INCH PIECES

2 TABLESPOONS VEGETABLE OIL

2 TEASPOONS GOOD-QUALITY CURRY POWDER

SALT

1 LIME

Preheat the oven to 350 degrees.

Toss the carrots with the oil, 1 teaspoon of the curry powder, and salt and place them in a roasting pan. Roast, stirring often, until almost tender, about 45 minutes. Sprinkle with the remaining curry powder and grate lime zest on top.

☞ Spices added before cooking have a very different effect from that of spices added afterward. Try a bit of roasted carrot before sprinkling the curry at the end, to see how the carrot and curry flavor have melded. The fresh addition wakes up what have become deeper, more latent flavors, and also brings a bit of heat.

THE FOUR RULES
OF FLAVOR

For Paracelsus, alchemy and the arts are one and the same:
"a baker is an alchemist when he bakes bread, the vintner
when he makes wine, the weaver in that he makes cloth."...
Alchemists were all those artisans who knew how
to separate the useful from the useless.

PAMELA H. SMITH, *THE BODY OF THE ARTISAN*

An ingredient doesn't start to become a dish until it's combined with other ingredients. But how do we choose them?

By default, the choice is usually made for us. Most of the combinations of ingredients we eat are part of a traditional canon: pork and beans mellowed and deepened with a dollop of molasses, spaghetti and meatballs in marinara sauce flavored with garlic and basil, roast turkey with a buttery stuffing redolent of onions, sage, and thyme. Even less traditional dishes quickly become stereotyped, like the ubiquitous beet and goat cheese salad. We know that some combinations work, because we've tasted them and they are delicious. They are what we eat in restaurants and at home; recipes for them are included in cookbook after cookbook, and your grandmother's and mother's subtly personalized versions of them—the parsnip and sprig of dill they always threw into their chicken soup, for example—have been passed down through the generations. But simply replicating these combinations doesn't make us understand how and why they work together. Nor does it help us to break out of them to develop our own distinctive style of cooking and eating.

Because food is so closely linked to what we feel and remember, it's easy to confuse what is familiar with what is good. To think originally about flavor, we have to learn to approach it fresh. The sensibility and vocabulary we've begun to develop for ingredients themselves points the way toward how to think about combining them. Instead of falling back on a CliffsNotes impression of an ingredient (peaches are sweet), we need to explore the full ranges of their attributes, with an attention to the way the nuances interact.

What allows great flavor to happen? As we say when people meet and have that special something that seems to transcend the boundaries of their isolated selves, it's chemistry. Literally. What's actually happening when ingredients take that quantum leap to exciting flavor is they are interacting on a molecular level in ways that heighten, layer, tame, and connect all the nuances of which they are composed, bonding together in a pleasurable way. We want the effect to be surprising, but as we shall see, it's a surprise that can be orchestrated.

Think about music. Although we experience it as romantic, intuitive, and emotional, music is actually governed by mathematical relationships—the structural arrangement of sound. In the same way, while dining is rooted in sensual pleasure and the social gratifications of sharing the table with loved ones, what underlies that experience is the molecular interaction of ingredients. It's not necessary to have a profound understanding of that underlying chemistry, but it is the basis of four broad rules we've distilled that govern how flavor works:

1. Similar ingredients need a contrasting flavor.
2. Contrasting ingredients need a unifying flavor.
3. Heavy flavors will need a lifting note.
4. Light flavors will need to be grounded.

It seems quite simple, but the essentials of flavor lie within these four basic principles. Mastering them will make you a better, freer, and more original cook.

Rule 1. Similar ingredients need a contrasting flavor.

Complementarity—combining ingredients that are similar in character, and similarly easygoing—is the path of least resistance. When we're in the mood for something homey—"comfort food"—we gravitate toward flavors that blend into each other: mac and cheese, for example, or rice and beans. There is something soothing about flavors that are close to each other, like two old friends who always get along.

It's not that such combinations can't be interesting, even sophisticated, in a subtle way. But in food, as in life, too much harmony breeds boredom. To avoid a pileup of blandness, some counterpoint is needed: something to punctuate the dish and give it shape. Even comfort food needs a dash of contrast to be interesting—at least a dash of cayenne in that mac and cheese.

Potatoes and leeks are one such cozy combination. The two ingredients share sweet, earthy, bland characteristics. It is a very common combination, but all too often a boring one.

Cooking methods, as we'll explore in chapter 7, can help bring out the differences. Cooking the potatoes in broth, pureeing them, and introducing cream and butter can bring out their sweetness and depth, and garnishing it with crisp fried leeks, perhaps enhanced by breading, can introduce complex browned flavors.

But we can also think about introducing a divergent flavor that acts as a counterpoint. Imagine that the potatoes and the leeks are, in fact, simply boiled in salted water and combined. What will create excitement?

Acidity suggests a direction—a salad, likely, with an acidic vinaigrette—but acidity is not a flavor per se. How will we compose that vinaigrette? Cured black olives beckon—salty, dark, fruity, and complex. So might oily, fishy cured anchovies and pungent green capers in brine. These intense flavors have the capacity to stand up to each other, and have facets that link together seamlessly. They also need to be used in restrained amounts—"buried," as we will discuss later. Treated in this manner, they will bind together to make a bold seasoning for a salad. Chopped parsley will add a fresh, grassy note. Lemon juice (for acidity and subtle flavor) and zest (a bright element to lift the deeper flavors), champagne vinegar (for neutral acidity), and fruity olive oil (to balance the acidity and bring both texture and flavors together) complete an easy but powerful vinaigrette. Contrast can be introduced into the base ingredients as well. A hearty lettuce like radicchio can provide not only a textural counterpoint to the softness of the potatoes and leeks, but also a pleasingly bitter foil for their sweetness.

Potato-Leek Salad with Radicchio and Olive-Anchovy Vinaigrette

SERVES 6

For the vinaigrette:

3 ANCHOVY FILLETS, CHOPPED

2 TABLESPOONS CHOPPED CURED BLACK OLIVES

1 TABLESPOON CHOPPED CAPERS

1 CLOVE GARLIC, FINELY GRATED (OPTIONAL)

2 TABLESPOONS FRESHLY SQUEEZED LEMON JUICE

FINELY GRATED ZEST OF 1 LEMON

4 TABLESPOONS CHAMPAGNE VINEGAR

4 TABLESPOONS FRUITY OLIVE OIL

SALT

For the salad:

1 POUND SMALL BOILING POTATOES

2–3 LEEKS, WHITE AND PALE GREEN PARTS ONLY, CLEANED

1 HEAD RADICCHIO

SALT AND FRESHLY GROUND BLACK PEPPER

Make the vinaigrette: In a small bowl, combine the anchovies, olives, and capers. Add the other ingredients a bit at a time, adjusting until you are pleased with the balance of flavors. Remember that the intensity of the dressing will be diffused a bit when it is combined with the salad ingredients. This part can be done up to a day ahead.

Make the salad: Simmer the potatoes in salted water until they are tender, then drain them. If they are larger than bite-size, cut them in half. Cut the leeks into bite-size pieces and simmer them in salted water as well, then drain them. Cut the radicchio into bite-size pieces. In a salad bowl, combine the potatoes, leeks, and radicchio with the vinaigrette and season with salt and black pepper to taste.

☞ Onions and garlic, as we'll discuss later, are foundational aromatics. Without the optional garlic, the vinaigrette is fresh and clean and bright. The garlic adds pungency and intensity, locking with the other flavors and heightening the umami. With or without the garlic, this also makes a wonderful sauce for fish.

Celery Root – Apple Salad with Creamy Mustard Dressing

SERVES 4 AS A MAIN COURSE, 6 AS A SIDE DISH

This is a variation on using a sharp ingredient to punctuate a salad. Celery root is a bulbous vegetable covered in a thick, brown, often furry skin. Underneath it is pure white, with a mellow celery flavor tinged with sweet, round, rootlike notes, a hint of potato and parsnip. The texture is crisp and snappy when fresh, tender when fully cooked. It's a very flexible ingredient, equally good raw, roasted, boiled, pureed, or simmered in broth.

Whether celery root is eaten raw or cooked, its flavor is vegetal and flat, and it is greatly improved by fruity and sharp notes. When it is raw and julienned, its appearance and texture are close to those of the apple, albeit less crunchy. Here we're playing with that likeness by introducing flavor contrast via a tart green apple. Still, the

> To make dry Mustard very pleasant in little Loaves or Cakes to carry in ones Pocket, or to keep dry for use at any time: Take two ounces of seamy [mustard], half an ounce of cinnamon, and beat them in a mortar very fine with a little vinegar, and honey, make a perfect paste of it, and make it into little cakes or loaves, dry them in the sun or in an oven, and when you would use them, dissolve half a loaf or cake with some vinegar, wine, or verjuyce.
>
> —ROBERT MAY, *The Accomplisht Cook* (1660)

combination needs a piquant note to lift it out of the mid-range and provide spark and excitement. Mustard provides that. A very old ingredient, mustard has long been relied upon in the West as well as the East to provide relief from gastronomic blandness——in the same way that it's still our standby to spike a deli sandwich. Here it is used in a dressing made creamy with mayonnaise and crème fraîche, the crème fraîche also providing a subtle tang that merges with the mustard's sharp, lifting acidity. The chives provide a fresh green note, and an oniony counterpoint.

1 CELERY ROOT

1 TART GREEN APPLE

2 TABLESPOONS MAYONNAISE

2 TABLESPOONS WHOLE-GRAIN MUSTARD

3 TABLESPOONS CRÈME FRAÎCHE

2 TABLESPOONS FRESHLY SQUEEZED LEMON JUICE

2 TABLESPOONS MINCED CHIVES

SALT AND FRESHLY GROUND BLACK PEPPER

Peel and julienne the celery root and apple. In a mixing bowl, whisk together the mayonnaise, mustard, crème fraîche, and lemon juice to taste. Add the celery root, apple, and chives. Toss to coat well, then season with salt and black pepper. Serve on its own or as an accompaniment to meat or game.

☞ Should you find this salad too demure, its flavors almost too balanced and harmonious, amp up the mustard and take your cue from it to pile on the heat: seeded and finely chopped jalapeño can add a sweet, round spiciness that locks with the mustard and punches through the other flavors.

Buttered Rice with Corn, Leeks, and Basil

SERVES 6-8 AS A SIDE DISH

Corn and leeks are natural bedfellows. When cooked, both are sweet, but the oniony character of the leeks wraps around the sunny, rather one-note sweetness of the corn, giving it dimension. Rice and butter will happily join them, melding together in a harmonious way. What can provide the needed contrast to keep this combination from mellowing into dullness? One possibility: the sharp green anisey punch of basil, especially when it is added right before serving, so that its perfume jumps off the plate.

2 CUPS SHORT-GRAIN RICE

3 CUPS CHICKEN STOCK

2 TABLESPOONS BUTTER

2–3 LEEKS, WHITE AND PALE GREEN PARTS ONLY, SLICED THINLY

2 CUPS FRESH CORN KERNELS

SALT

GENEROUS HANDFUL OF FRESH BASIL LEAVES

Simmer the rice in the chicken stock over low heat, covered, until it is tender; there should be just a little liquid left unabsorbed. In a separate pan, melt the butter over medium heat. Add the leeks and sweat them until they are tender. Then add the corn kernels and cook, stirring often, until the vegetables are moist and barely tender, about 2 minutes—you don't want them to lose their freshness. Add this mixture to the rice, stir to combine, and season to taste with salt. Just before serving, top with a scattering of torn fresh basil leaves. Here, as in the previous recipe, feel free to keep layering on contrapuntal notes: Some lime zest over the basil, lifting the greenness even higher. Some chili or black pepper for spice.

☞ Try replacing the chicken stock with corn stock. (Stock is just flavored water—you can flavor it with anything. You can make a spice stock or an onion stock, as we will see later.) Put the cobs in a pot, cover them with water, and simmer for an hour. Taste. If you want the stock to be more intense, you can simmer it longer, or remove the cobs and boil it down to concentrate the flavor.

A COUNTERPOINT TO SIMILAR FLAVORS may be introduced in the form of a garnish or condiment, like the salty and spicy condiments commonly used to season the Asian rice porridge known as congee. Relishes and pickles function this way as well. This is a very old method of creating flavor at the table. In "The Pleasures of Consumption: The Birth of Medieval Islamic Cuisine," H. D. Miller notes that in medieval Arab cuisine, simple dishes such as roasted meats were typically accompanied by a wide range of foods pickled in spiced vinegar: cucumbers, eggplants, turnips, green beans, mint leaves, locusts, small fish.

Grilled Pork with Pickled Peppers

Sometimes a single ingredient contains multitudes. Pork is one such ingredient. The chop from a well-raised animal grilled to toothsome pinkness, a thick layer of fat banding the perimeter, has umami, sweetness, and a subtle gaminess, with nutty and browned notes. Pork can be so rich, however, that it benefits from a punctuating element. This can be a dollop of grainy mustard, an accompaniment of fermented cabbage like sauerkraut or kimchee, or, as in this recipe, a condiment of sweet and hot peppers. The acidity and heat leaven the heavy meatiness of the pork.

The recipe calls for red and yellow peppers, which are easily found in supermarkets, but if you have access to farmers' markets at the height of summer, use any varieties you can find, and make it as spicy or as mild as you want. You can also multiply the recipe to make a big batch to be canned—a bit of summer for the dark winter months to come.

1 CUP SWEET RED AND YELLOW PEPPERS CUT INTO 1-INCH DICE

1 SERRANO CHILI, THINLY SLICED

¾ CUP WHITE WINE VINEGAR

1 CUP WATER

1 TABLESPOON SALT

1 TABLESPOON SUGAR

4 THICK-CUT PORK CHOPS

SALT

Put the peppers and chili in a nonreactive bowl. In a small saucepan, bring the vinegar, water, salt, and sugar to a boil and pour over the peppers. Let cool to room temperature and refrigerate for at least a day so the flavors can meld.

Prepare a grill. Season the pork chops with salt, and grill, turning often, until the outside is browned and the interior is cooked but still pink. Let rest for 5 minutes and then serve with a spoonful of drained peppers on top.

☞ In the winter you can pickle chopped root vegetables this way. If you want the pickles more sweet or more sour, adjust the salt and sugar according to your taste when you first make the brine.

> *Rule 2. Contrasting ingredients need a unifying flavor.*

All ingredients have their differences, but contrasting ingredients are different in a fundamental way, the way that red is different from blue—or even more, the way that pale pink is different from a deep, saturated teal. These are ingredients that differ radically in character, shape, texture, and intensity: fish and flower, root vegetable and spice.

Beginning by juxtaposing two contrasting ingredients automatically creates more dynamic and interesting possibilities in the imagination. It generates excitement, aliveness, a yin and yang. The tension between the ingredients translates to a liveliness in the mouth. Sometimes divergent flavors can fuse together to create a new flavor, the phenomenon of locking that we'll investigate in chapter 6. But almost always, contrasting flavors need additional ingredients to bridge the distance between them.

Cauliflower with Cumin Seed and Browned Butter

The bridge between contrasting flavors can be a very simple one. Cauliflower is a natural partner to strong flavors. It is sweetest and most flavorful when fully cooked, but boiling turns it submissive, requiring the rescue of a strong sauce or vinaigrette, and also brings out a sulfurous quality. Parboiling and then roasting, grilling, or sautéing it, however, brings out its sweet and savory notes and leaves it able to hold its own. Here its partner is cumin, with its warm, spicy, penetrating, almost sweaty and slightly bitter taste and aroma. When the cumin is dry roasted, the sweaty aspects recede and its earthy, spicy warmth comes forward. With its complex flavor, it wouldn't combine well with simple melted butter. But when the butter browns—or rather, when the water in the butter burns off and the milk solids fry in the remaining fat—rich and meaty aromatics are created, much as when you brown meat, and the browned butter stands up well to the strong spice.

The flavors are all very rich and intense, but the brown butter unites the cauliflower and cumin, and lime zest cements the bridge, its fresh green note locking with both ingredients and balancing the flavors. Laid out on a bed of tender lentils, perhaps, this is a dish hearty and interesting enough to serve as the main element of a dinner.

SALT

1 MEDIUM HEAD CAULIFLOWER

2 TEASPOONS CUMIN SEED

4 TABLESPOONS BUTTER

1 LIME

Bring a large pot of water to a boil and salt it well. While the water is heating, cut the cauliflower into quarters, leaving the florets joined at the base. Add the cauliflower to the water and boil until half tender, 8 to 10 minutes. Drain.

While the cauliflower is cooking, toast the cumin in a small dry skillet and then grind it in a spice grinder or with a mortar and pestle.

In a large skillet, melt the butter over medium heat. Add the cauliflower and cook it, turning often, until the quarters are well browned on each side and the butter has developed a deep, nutty aroma. Sprinkle the cumin over the browned cauliflower and cook a few minutes more, seasoning lightly with salt. Pay attention to how the aroma of the cumin rises and then merges with that of the browned butter and the vegetal tones of the cauliflower.

Plate the cauliflower, drizzle the butter over it, and zest a little lime over the top.

☞ This is a dish that needs acidity for balance. The lime zest provides spark, performing almost like an herb. But the deep and savory flavors could use another counterpoint. Lime juice on its own will diminish the savory qualities you have worked so hard to build: acidity diminishes umami and dilutes concentration. Instead, you could add lime juice to yogurt for a bright, sharp dipping sauce, which will provide relief without undermining the flavor of the cauliflower.

Sometimes contrasting ingredients are very far apart on the flavor spectrum. Lamb and anchovies don't seem to have much at all in common. Moreover, each has a highly assertive character of its own. Yet as cuisines all over the world reflect, flavors of the ocean, which are usually sharp and saline, can be a gorgeous complement to the deep, rich, round flavors of the earth. Lamb and anchovy, strong as they are, can make a fabulous contrast. For example, you might grill lamb chops—or a boneless leg of lamb—to pair with the Potato-Leek Salad with Radicchio and Olive-Anchovy Vinaigrette (page 63). The comforting blandness of the potatoes and leeks creates a very long bridge that unites these two strong elements. Cooking method plays a role in modulating them as well: grilling tames the distinctive gaminess of the lamb, and mincing the anchovy and mixing it into the vinaigrette diffuses its intensity, so it becomes a nuanced seasoning rather than a dominant flavor.

The bridge between highly contrasting ingredients doesn't always have to be a bland one, as this variation on lamb with anchovy demonstrates. Here a vinaigrette, made deeper and more interesting by the addition of garum (a fermented anchovy sauce that dates at least to ancient Rome), is the bridge.

Grilled Lamb with Fermented Anchovy Sauce and Bitter Greens

SERVES 4

The lamb is, again, grilled in this recipe, but the anchovy is in the form of the fermented fish sauce called garum. Produced by fermenting anchovies with salt, garum is like a more mellow and elegant version of fish sauce. Like any fishy ingredient, garum benefits from the bright, gentle acidity of lemon, which freshens it. Or, to put it another way, a touch of garum turns a simple lemon vinaigrette deeply flavored and far more interesting, adding subtle fermented, gamey notes. Combined with a bit of freshly grated raw garlic and a little red wine vinegar, this makes a vinaigrette that can both pull in and stand up to the deep, savory lamb. The bitter greens provide contrast.

8 LOIN OR RIB LAMB CHOPS

2 TABLESPOONS OLIVE OIL PLUS MORE FOR GRILLING

SALT AND FRESHLY GROUND BLACK PEPPER

2 TABLESPOONS FRESHLY SQUEEZED LEMON JUICE

2 TABLESPOONS RED WINE VINEGAR

2 TEASPOONS GARUM

1 SMALL CLOVE GARLIC, GRATED WITH A MICROPLANE OR FINELY CHOPPED

A MIX OF BITTER GREENS—ENDIVE, RADICCHIO, DANDELION, FRISÉE

Rub the lamb chops with oil and season with salt and black pepper. Grill over charcoal or in a pan, turning often, until well browned and cooked to the degree of doneness you desire.

While the lamb is cooking, make the vinaigrette: Combine the lemon juice, vinegar, garum, garlic, 2 tablespoons olive oil, and salt to taste. After the lamb is finished, dress the greens with the vinaigrette and serve next to the lamb.

☞ This vinaigrette, with the addition of onions that have been grilled and minced, becomes a relish with which you can sauce the lamb itself. If you can't find garum, a good fish sauce works well.

THE FLAVOR SHAPE of an ingredient says a lot about whether or not it will make a good bridge. Strong flavors are usually pointy—think of spikes or being stuck with a pin, which produces an intense, immediate sensation. Bridges are usually flat and wide, creating a calm, comforting feeling. That could mean bland, like potatoes, or sweet, like dates. If blue cheese is sharp and has a shape that rises like a mountaintop, dates are a low, flat, accommodating plain, on which different flavors—like blue cheese and bacon—can come together.

Medjool Dates Baked with Bacon and Gorgonzola Dolce

Dates can be eaten raw, but they are usually sold dried. They have a high proportion of sugar and no saltiness or acidity, which makes them a good link between two intense flavors. Here they hold bacon and blue cheese in balance. It can work with any dates and any blue cheese, but the best marriage depends on considering the particular facets of each. Medjool dates have a caramel-like flavor with tones of honey and a soft, creamy texture. Their relatively large size makes them ideal for stuffing, especially when the pit is removed. Gorgonzola dolce is one of the mildest and creamiest of blue cheeses, with the kind of rich, honeyed sweetness associated with mascarpone. The sweetness accommodates the bacon, but the cheese has enough umami from its fermentation not to be overwhelmed by its smokiness.

THINLY SLICED SMOKED BACON

MEDJOOL DATES

GORGONZOLA DOLCE

HONEY FOR DRIZZLING (OPTIONAL)

Preheat the oven to 350 degrees.

Lay bacon slices on a sheet pan and bake until cooked through but still pliable. Drain and cool to room temperature.

Halve the dates, remove the pits with a small knife, and stuff the cavities with as much Gorgonzola as will comfortably fit. Close the dates around the Gorgonzola and wrap each with a piece of bacon, trimmed to overlap by half an inch, and secure it with a toothpick. If not using immediately, refrigerate.

Preheat the oven to 400 degrees. Spread the dates in a baking pan and roast until the outside is crisp, about 7 to 8 minutes. Let cool slightly before serving, with a drizzle of honey on top, if desired, to bring extra sweetness and soften the flavor.

☞ Proportion is especially important with strong flavors. The amount of blue cheese is naturally limited by the cavity of the date. The bacon pieces should be no wider than the dates. If the bacon is too thick it won't get crisp, and if there is too much of it the smoky fat will dominate.

As you become more comfortable combining contrasting ingredients, you can start introducing greater complexity into your dishes. Rather than think in terms of ingredient plus ingredient, bridged by a peacemaker, you can begin to think in terms of layers. Instead of dancing a two-step, in other words, you can start doing the salsa. Many non-European cuisines embody just such an approach, with a careful orchestration of multiple strong flavors playing out on top of a bland base, layered and proportioned in such a way as to create a balanced and delicious whole. Many curries are constructed like this, using some combination of ingredients such as cumin, coriander, fenugreek, ginger, turmeric, and cardamom, among many others, to create a harmonious layering of disparate flavors.

Chicken and Eggplant Cooked with Aromatic Spices

SERVES 6

- **2** TABLESPOONS VEGETABLE OIL
- **2** TABLESPOONS BUTTER
- **6** CHICKEN THIGHS
- SALT AND FRESHLY GROUND BLACK PEPPER
- **2** TEASPOONS TURMERIC
- **¼** TEASPOON CHILI FLAKES
- **2** TEASPOONS GROUND CUMIN
- **1** TEASPOON GROUND FENUGREEK
- **1** TEASPOON GROUND FENNEL SEED
- **1** TEASPOON GROUND CARDAMOM
- **3** CLOVES GARLIC, MINCED
- **2** TEASPOONS MINCED FRESH GINGER
- **3** JAPANESE EGGPLANTS, CUT INTO CHUNKS
- **3** CUPS CHICKEN STOCK
- FRESHLY SQUEEZED LIME JUICE
- FRESHLY GRATED LIME ZEST
- OPTIONAL FOR GARNISH: MINCED SCALLIONS, CILANTRO, BASIL, MINT

In a large sauté pan, heat the oil and butter over medium-high heat until the butter is bubbling. Season the chicken with salt and black pepper and add it to the pan. Cook, turning often, until the chicken is browned on all sides. Remove the chicken to a plate.

There should be a lot of fat left in the pan. Remove the pan from the heat to let it cool a bit, then add the spices and cook for 5 minutes at low heat, stirring often. Add the garlic and ginger and cook for 1 minute, until aromatic, then add the eggplant chunks and stir until completely coated in the fat and spices.

Add the chicken stock and return the chicken to the pot. Simmer the chicken over low heat, covered, until it is tender, about 1 hour.

Taste the liquid; it should be seamless. Rebalance as needed, then add the lime juice and zest for the desired brightness. Serve over rice, sprinkled with scallions, cilantro, basil, mint, or any combination of them as desired for additional lift.

☞ Long cooking is a basic method for melding disparate flavors into one, and this recipe can be made with almost infinite variations to arrive at new flavors. Making this the day before and letting it rest in the refrigerator overnight will only improve it.

TOP, MIDDLE, AND BASE NOTES

Before we go into the third and fourth rules, it's helpful to revisit the perfume studio, with its neat categories of top, middle, and base notes. As discussed, ingredients aren't so neatly categorized in the kitchen, but they play different roles in different dishes, depending on the method of preparation and what company they're keeping. As in a musical chord, the base notes are the ingredients that play an anchoring role, giving the dish depth and resonance. The top notes—most often herbs and spices—greet nose and tongue first, lift the dish, and give it lightness and delicacy. And the middle notes connect top and bottom and add richness. Balance is achieved when all three kinds of notes are present in correct proportion. And as in music, simply combining notes at different registers doesn't guarantee a pleasing whole. Some combinations create harmony, and some create dissonance. Creating flavor in a controlled way comes from knowing the characteristics of your ingredients intimately, being in touch with your own preferences, and learning how to choose and orchestrate the ingredients that will best match the hungers of your imagination.

> Rule 3. Heavy flavors need a lifting note.

Base notes in food are played by the heaviest, densest flavors in a dish—those that linger in the mouth. They are often given the biggest roles in the dish, in terms of real estate, but not necessarily those that make it most distinctive or memorable. Think of the actor with the most lines versus the supporting players who steal the show—or more precisely, whose scenes with the principals are what make the ensemble a smash hit. Most of what we think of as comfort foods—pot roasts and pudding, for example—are heavy on base notes.

Although these flavors are soothing, they are not very exciting. As with similar ingredients, heavy flavors need a lift—a parsley-lemon gremolata for that pot roast, a pinch of citrus zest for the pudding. Bases and middle notes need a top note to complete the chord.

Braised Greens with Lemon and Olive Oil

The part of base isn't always played by a heavy protein or starch; again, it's relative. We think of greens as bright; and when they're fresh, they are. But cooked greens, save for sorrel, are flat, and served as a side dish, they play base, whether it's bitter, earthy kale; very bitter broccoli rabe or dandelion greens; sweeter and slightly metallic Swiss chard; or sweet and spicy mustard greens. Without exception, acids like lemon and vinegar make cooked greens tastier, but the exact kind of acid that might work best and the effect it will create vary with the greens.

It is important to cook the greens until tender, especially the sturdiest greens like kale and chard, not only for an agreeable texture but also to turn their grassy, vegetal notes sweet. The other general rule is to not use too much water—to concentrate rather than dilute the flavor. Keep just enough water in the pan to boil or steam the greens.

Along with an acidic lift, almost any seasoning will work with braised greens. A few standbys are onion and garlic, sautéed first until aromatic. Among many possible spices, chili flakes add a focused brightness and edge, and fish sauce brings a transformative funkiness. Use just a little, and it will marry beautifully with the lemon and olive oil. Braised greens also make a terrific middle note in meat and fish stews and a welcome partner to root vegetables and tomatoes.

GREENS, WASHED AND CUT INTO 1-INCH STRIPS

SALT AND FRESHLY GROUND BLACK PEPPER

FRESHLY SQUEEZED LEMON JUICE

FRUITY OLIVE OIL

OPTIONAL: GARLIC, MINCED ONIONS, CHILI FLAKES OR OTHER SPICES,
FISH SAUCE

For the basic preparation, place the greens in a nonreactive pot, add a small amount of water and some salt (not too much, as the liquid will reduce), and cook over medium heat, covered, until the greens are completely tender. Stir occasionally, checking to make sure they aren't burning, and add a little more water if necessary. There should be only a small amount of water left in the pot when the greens are cooked.

Drain the remaining liquid, then add lemon juice and olive oil and season with salt and black pepper. As a general rule, the more bitter the greens, the more lemon it will take to tame the bitterness.

☞ If using garlic, onion, chili flakes, or other spices, build their cooking into the dish. Sauté them in a couple of tablespoons of olive oil until aromatic but not browned, then add the greens and water. That way they contribute their flavor not only directly but also to the oil, which will in turn flavor the cooking liquid. Fish sauce can be added either at the beginning, so it will meld in, or at the end, so it will remain more distinct, depending on the effect you want.

Farro Salad with Feta and Mint

The flavor of grains is usually flat, and salads based on them require a lift. One way to broaden the middle is to add another ingredient with sweet and flat flavor, like peppers or cucumber. This gives more of a base for the acidity and herbs required to energize the combination.

In this case, young artichokes are shaved raw, adding body, texture, and vegetal sweetness. The earthy/fresh facets of the artichoke connect with the earthy/sweet flavor of the farro. To lift the combination, a simple dressing of rice wine vinegar and fruity olive oil, with lots of sliced mint. Almonds provide additional texture, and their nutty flavor plays off the nutty facets of the farro.

Crumbled feta adds a lightly sour tang and soft creaminess, a respite from all the hard textures.

2 CUPS COOKED FARRO

8 YOUNG ARTICHOKES, PEELED AND SHAVED

4 TABLESPOONS RICE WINE VINEGAR

4 TABLESPOONS FRUITY OLIVE OIL

4 TABLESPOONS TOASTED AND CHOPPED ALMONDS

4 TABLESPOONS CRUMBLED FETA

½ CUP SHAVED FENNEL

SALT AND FRESHLY GROUND BLACK PEPPER

10 LEAVES MINT, SLICED

Combine all the ingredients except the mint in a mixing bowl. Toss, taste, and adjust seasonings. Divide among serving dishes. Shower with mint.

When making salad, it's sometimes hard to know whether it's better to add an ingredient when tossing or to sprinkle it on top at the end. In this case, the feta tossed with the salad will break down somewhat and make the dressing a little creamy, providing a binder for the crunchy ingredients. The mint, however, should be added at the end, so it will give more of a spark.

Meatballs with Penne and Citrus-Herb Sauce

SERVES 4-6

Meatballs are a classic comfort food, and ground meat is unquestionably a base note in need of a lift. The traditional way of proceeding might be by introducing the sweetness and acidity of the tomatoes in the middle register, maybe calling out their lifting acidity with a splash of balsamic, brightening the whole with a scattering of fresh basil, and adding richness, tang, and depth with a sprinkle of Parmesan cheese. But looking at the dish without the lens of tradition might allow you to go in a different direction. Instead of basil, why not look for a lift from a mixture of different citrus zests and herbs?

For the citrus-herb sauce:

2 TEASPOONS FRESHLY GRATED ORANGE ZEST

2 TEASPOONS FRESHLY GRATED LEMON ZEST

2 TEASPOONS FRESHLY GRATED LIME ZEST

1 TABLESPOON MINCED PARSLEY

1 TABLESPOON MINCED CHIVES

1 TABLESPOON MINCED TARRAGON

OLIVE OIL

FRESHLY SQUEEZED LEMON JUICE

SALT

For the meatballs:

½ POUND GROUND PORK

½ POUND GROUND CHICKEN

1 CUP WHOLE-MILK RICOTTA

½ CUP BREADCRUMBS

2 TEASPOONS SALT

SEVERAL GRINDS BLACK PEPPER

VEGETABLE OIL

1 QUART SIMPLE TOMATO SAUCE, HOMEMADE OR PREPARED

To complete the dish:

1 POUND PENNE OR OTHER TUBULAR PASTA

Make the citrus-herb sauce: In a small bowl, combine all the ingredients for the sauce. Taste and adjust as desired.

Make the meatballs: Combine the ground pork and chicken, ricotta, breadcrumbs, and salt and black pepper. Mix well with your hands. Form the mixture into small balls. Place a wide sauté pan over high heat and film with the oil. Add the meatballs and sear, shaking the pan to brown them on all sides. Add the tomato sauce and a splash of water. Cover and simmer for 20 minutes, or until the meatballs are cooked through.

Bring a large pot of water to a boil and salt it well. Add the pasta and cook until it is al dente. Drain the pasta and mix it with the meatballs and tomato sauce. Divide the pasta among serving plates and drizzle with the citrus-herb sauce.

☞ Try this with the Tomato-Miso Sauce (page 187).

Chocolate Pots de Crème with Ginger Cream and Rose-Cardamom Sugar

SERVES 6-8

Chocolate is an ingredient that sits between sweet and savory, ready to go in either direction. The earthy, bitter, fruity tones of unsweetened chocolate diminish with each addition of sugar and fat, until it is finally tamed into a dessert. But a little bit of salt brings back some of its savory elements. In this recipe, the earthy bitterness of coffee is lifted by whipped cream infused with ginger and lemon zest. Croutons baked with sugar instead of salt are a bit of a sleight of hand: the texture of a savory element but the flavor of a sweet one. The sugar is ground with dried rose petals and a pinch of cardamom, which adds a light floral-spice component.

For the pots de crème:

12 EGG YOLKS

6 TABLESPOONS SUGAR

1½ CUPS WHOLE MILK

⅓ TEASPOON SALT

2 CUPS HEAVY CREAM

⅔ CUP STRONG COFFEE OR ESPRESSO

15 OUNCES 64—70 PERCENT CACAO CHOCOLATE

For the rose-cardamom sugar:

¼ CUP ROSE PETALS (AVAILABLE IN MIDDLE EASTERN OR SOUTH ASIAN GROCERIES, OR DRY YOUR OWN)

¼ TEASPOON GROUND CARDAMOM

½ CUP SUGAR

For the croutons:

1½ CUPS DICED *PAIN DE MIE* OR BRIOCHE

4 TABLESPOONS BUTTER, MELTED

2 TABLESPOONS ROSE-CARDAMOM SUGAR

For the ginger whipped cream:

1½ CUPS HEAVY CREAM

1 TABLESPOON FINELY GRATED GINGER

4 TABLESPOONS SUGAR

FRESHLY GRATED ZEST OF 1 LEMON

Preheat the oven to 300 degrees.

Make the custard: In a mixing bowl, whisk together the egg yolks and sugar until well combined. Whisk in the milk and salt. In a medium saucepan, heat the cream and coffee until almost at a simmer. Remove from the heat and add the chocolate. Stir until the chocolate is melted. Let cool slightly. Add the chocolate mixture to the egg mixture, and whisk to combine. Divide the mixture evenly among ovenproof custard cups and place them in a large baking pan. Add boiling water to the pan to the depth of an inch or two and bake until the center of each custard is set around the edges but still jiggly in the middle, about 30 minutes. Let cool and refrigerate for at least 3 hours.

Make the rose-cardamom sugar: Grind the rose petals in a spice grinder until powdered. Add the cardamom and sugar and grind until completely incorporated. The sugar should have a nice pink color.

Make the croutons: Toss the *pain de mie* or brioche with the melted butter. Spread on a baking sheet and bake, turning often, until the croutons are golden brown. Add the rose-cardamom sugar and toss to coat completely. Let cool.

Make the whipped cream: Combine the ingredients and whip to soft peaks.

When ready to serve, garnish each cup of custard with a dollop of whipped cream and several croutons.

☞ You can make scented sugar or salt by grinding any kind of herb or spice with sugar or salt as in this recipe. If using herbs, pass them through a strainer basket after grinding to remove any big pieces.

Oleo-saccharum is a sugared oil from oleo (oil) and saccharum (sugar). Such oils were used in the nineteenth century to add aroma and flavor to drinks. Here is an old recipe for an oleo-saccharum from the 1869 book *Cooling Cups and Dainty Drinks* by William Terrington:

In everything in which lemons are used the peel should be cut very thin, by reason that the flavour and scent, which constitute its most valuable properties, reside in minute cells, close to the surface of the fruit, so, by slicing it very thin, the whole of the minute receptacles are cut through, and double the quantity of the oil is obtained; or the outer rind may be rubbed with a lump of sugar, which, as it breaks the delicate vessels, absorbs the ambrosial essence. To make the sugared essence (or oleo-saccharum), either pursue the above method, and as the sugar is impregnated with the essence, scrape it off with a knife from the lump, or peel some lemons very thin, and pound the peel into a stiff dry paste in a marble mortar, with sufficient sugar, and preserve it for use, closely pressed in a tightly covered jar.

Rule 4: Light flavors need to be grounded.

Sometimes what your ingredients need isn't lift, but depth. Light, bright flavors can be potent and forward in an appealing way, but they need a platform on which to dance. Again, registers are relative: a vegetable or even a spice can provide the grounding that's needed for a delicate combination. Conversely, a vegetable or grain can seem insubstantial or unsatisfying without a deeper note.

Balance, of course, is relative, and dependent on the unique character of your ingredients, so it can't be precisely measured or codified but instead depends on your judgments and adjustments in the moment. Generally, though, the use of fat, earthy flavors and the addition of fermented or umami-rich ingredients are good ways to deepen flavors in order to create a balanced combination. Also, light flavors can be concentrated—through reduction, which removes water, or through long cooking or the brief application of heat, which can make them more profound.

Peas with Egg Yolk Sauce and Tarragon

Spring vegetables are almost all light and sweet, the first delicate yawn of farms just waking from winter's slumber. Peas at the height of the season, in particular, benefit from a quick dalliance with butter in a sauté pan, a little salt, and nothing else. But that's not really a flavor combination. To turn lightly cooked peas into a composed dish, they need a grounding element—in this case, egg yolks, in the form of a light but rich béarnaise, spiked with vinegar and tarragon.

1 CUP PLUS 1 TABLESPOON BUTTER

2 TEASPOONS CHAMPAGNE VINEGAR

2 EGG YOLKS

1 TEASPOON FRESHLY SQUEEZED LEMON JUICE

SALT

2 TABLESPOONS WARM WATER

1 CUP SHELLED ENGLISH PEAS

1 CUP SNAP PEAS, TIPS AND STRINGS REMOVED, CUT IN HALF

1 TABLESPOON MINCED FRESH TARRAGON

PEA SHOOTS (OPTIONAL)

Place 1 cup of the butter in a sauté pan over medium heat and cook until it is lightly browned, being careful not to let it burn.

Put the vinegar, egg yolks, and lemon juice in a blender with a little salt. Add the warm water. Blend well, then, on low speed, drizzle in the warm browned butter. Don't overblend; stop when the butter is absorbed. Adjust seasonings.

Put the remaining 1 tablespoon butter in the sauté pan and place it over medium heat. Add the peas and snap peas and some salt. Cover and cook for 2 to 3 minutes, until the peas are just cooked through. Then stir in the tarragon.

Divide the peas among serving bowls and pour the sauce over and around them. Garnish with pea shoots if you have them.

☞ This is a very focused dish, but the addition of other green and white spring vegetables—asparagus, new onions, fava beans—would be welcome.

Winter Vegetables with Charred Onion Broth and Olive Oil

Vegetables are thought of mostly as light, but this is a versatile vegetable broth with the depth of a meat sauce. Onions are burned over high heat and then simmered until sweet and rich. The bitterness balances the sweetness and gives it dimension, making it smoky and complex. The broth is infused with aromatic lemongrass to lift it, seasoned with sherry vinegar, and served with winter vegetables that have been poached and grilled, and a drizzle of olive oil. The light charring of the vegetables heightens the intensity and connects them to the browned facets of the broth.

In this recipe, the concentration of the broth is everything. Too much water and it is diluted, too little and it becomes dense and muddy. It should be both powerful and balanced, to enrich the vegetables without overwhelming them. The concentration and *sweetness also balance the vinegar, which plays against the olive oil in a sort of warm vinaigrette.*

4 MEDIUM ONIONS, HALVED

2 QUARTS WATER

1 STALK LEMONGRASS

SALT

SHERRY VINEGAR

2 CUPS BROCCOLI FLORETS

2 CUPS CAULIFLOWER FLORETS

12 NEW POTATOES

1 BULB FENNEL, THICKLY SLICED

2 LEEKS, WHITE AND PALE GREEN PARTS ONLY, THICKLY SLICED

4 TABLESPOONS PURE OLIVE OIL

4 TABLESPOONS FRUITY, FULL-FLAVORED OLIVE OIL FOR GARNISH

WHOLE BLACK PEPPERCORNS

In a very hot cast-iron skillet or on a griddle, or over an open flame, cook the onions, cut side down, until blackened on at least half the surface. Transfer to a heavy pot. Add the water, bring to a boil, reduce the heat to very low, and simmer for at least 2 hours, until the broth is flavorful. Strain the broth and return it to the pot. Add the lemongrass and steep for several minutes, to desired intensity. Season with salt and sherry vinegar—the broth should have a gentle acidity, just enough to balance the sweetness.

Bring a pot of water to a boil, salt, and add the remaining vegetables. Simmer until tender, then drain and reserve.

To serve, return the broth to a simmer. Toss the vegetables with the pure olive oil and salt, and char in a hot cast-iron pan or on a griddle until lightly browned. Divide them among serving bowls and add about a ladle-ful of the broth. Drizzle with the fruity olive oil and grind a bit of black pepper over each.

☞ This recipe uses two kinds of olive oil, to very different effect. The pure, not very flavorful oil is just a vehicle for the charring of the vegetables. For the fruity olive oil—which makes a good dish great—try to find a freshly pressed oil. Because the solids have not settled out of the oil, it is more intensely aromatic and volatile.

Yogurt-Herb Sauce

Fat and texture can help ground a recipe. Here, a mixture of herbs, each of them a light flavor, is grounded with yogurt and brightened with lemon juice to make a sauce that works on everything from rice and meats to stews and grains. This is an all-purpose recipe, a template that can go in many different directions, with different combinations and proportions of herbs.

½ CUP MINCED PARSLEY

½ CUP MINCED CHERVIL

½ CUP MINCED CHIVES

½ CUP MINCED CILANTRO

½ CUP MINCED MINT

2 CUPS PLAIN YOGURT

2 TABLESPOONS FRESHLY SQUEEZED LEMON JUICE

1 TABLESPOON OLIVE OIL

SALT

Place all the ingredients in the pitcher of a blender and blend until the sauce is smooth and bright green. Adjust the lemon juice and salt.

☞ You can make this with any combination of sweet herbs, whatever you have on hand. In order to accommodate differing intensities of herbs—freshly picked mint versus week-old parsley—vary the amount of yogurt: with less flavorful herbs, use less yogurt; with intensely flavored herbs, use more.

Five

THE FLAVOR COMPASS

Though we tend to think of a particular herb or spice as having its own distinctive flavor, it too is always a composite of several different aroma compounds. Sometimes one of those compounds predominates and provides the main character—as in cloves, cinnamon, anise, thyme—but often it's the mixture that creates the character, and that makes a spice well suited to serve as a unifying bridge among several different ingredients. Coriander seed, for example, is simultaneously flowery and lemony; bay leaf combines eucalyptus, clove, pine, and flowery facets. It can be fascinating and useful to taste spices analytically, trying to perceive the separate components and how their flavors are built.

HAROLD McGEE, *ON FOOD AND COOKING*

Although every element of a dish contributes to its unique flavor, the ingredients that as a group contribute most—and most gloriously—are indisputably those that come from the parts of plants that are richest in essential oils: the rinds of citrus, the petals and stamens of a few edible flowers, the leaves we use as herbs, and the bark, roots, and seeds that we call spices. These ingredients typically take up very little real estate on the plate, but their impact is outsize. They are often the last ingredients to be added to a dish and the first that your nose and taste buds perceive. Their aromas are not just what draws a diner toward a dish, but those that first allow the cook to imagine what a dish will taste like when those ingredients are added to it. And because these ingredients are so extravagantly wealthy in flavor molecules, they are capable of producing transformative fireworks when they meet.

Yet we tend to underestimate the importance of these ingredients, and their potential. We rarely plan a dish around them, partly because—unlike, say, the beautiful pile of green beans at the farmers' market, or the sublimely fresh swordfish that calls to you at the fishmonger's—these ingredients don't tend to nominate themselves for star billing. When we do think of them, we tend to consider them primarily as afterthoughts, garnishes, add-ons.

Many of these ingredients you already have on hand in your pantry, in the form of little jars of spices. You may even have a window box of fresh herbs, or some bottles of edible essential oils. Some days, having that kind of abundance on hand is like a security blanket that makes you feel that you are ready to whip up something delicious at a moment's notice. But at other times the choice of what to work with can be paralyzing, especially when you are trying to break free of recipes, and of your tried-and-true—and tired—routines, in hopes of developing a more personal and delicious way of cooking.

In our passion to get you more deeply immersed in these highly flavorful ingredients, we've evolved a novel way of categorizing them—but not for the reason you might expect. Our aim is not to show you how to pick "one from column A, one from column B," but rather, by exploring the range of aromatic ingredients within a given family, to heighten your awareness of the particular qualities of each, and its unique potential as part of an ensemble.

To help perfume students think in this structured way, Mandy developed a perfume "wheel" that groups aromatics into families that share similar characteristics—floral, woody, spicy, and so on. Taking this idea into the kitchen, we have created a flavor "compass" meant to help guide you in the right direction by revealing both the similarities and differences among members of each major group. Our aim is to help you develop the attunement and sensual intelligence—and imagination—that will allow you to make connections in an easy, joyful, and fruitful way.

THE FOUR DIRECTIONS OF FLAVOR

Spices come from seeds, roots, and bark, often from plants grown in faraway places, giving them their tinge of the exotic. They are the richest and most intense family of flavors on the compass, vivifying neighboring ingredients and not only introducing new flavors to the mix but also highlighting those that are already present. They offer the greatest potential for creating contrast. In the perfumer's schema, most spices are middle notes. With their intense and distinct identities, they are usually dosed in modest amounts. Most are none the worse for being dried—and sometimes the more intense for it. They can be used raw, but they typically benefit from being mellowed, diluted, and sometimes amplified through cooking.

Herbs are aromatic leaves, and (with some exceptions) their volatile compounds diminish quickly after picking, so they don't dry flavorfully. Therefore, they are usually best when fresh, and their availability in that form sometimes depends on what is grown nearby. Unlike most spices, with their aura of the faraway and the exotic, herbs tend to remind us of gardens and home, where some of us indeed grow them. Their flavor is softer and lighter than that of spices. They are closer together—more similar—as a group, largely because of the considerable overlap of their most common aromatic molecules. Their family resemblances can make them easy to combine.

Many are bright top notes, best when eaten raw (think mint or basil), while others (rosemary and sage) strike a rich middle note and are better cooked.

Citruses, as we know from everyday experience, are fragrant in all their parts: flowers, rind, and juice—even (as with Makrut limes, popular in Thai cooking) in their leaves. But the heaviest concentration of their essential oils is found in small ductless glands in the outer portion of the peel. Given the difference in intensity, we employ the juice and peel in different ways. But even when citrus juice is expressed, a small amount of the oil from the peel ends up in the juice, heightening its flavor.

Citruses provide brightness, and they knit other flavors together. They are especially adept at bridging the distance between herbs, and between herbs and spices. The citruses are fairly close aromatically, but they have useful differences. For example, orange inclines toward sweetness, lemon toward tartness, and lime is the most tart of all, with an almost herbal freshness as well.

Flowers are used in cooking by many cultures around the world in many forms—fresh, dried, or distilled as essences to add unique flavors to food. Since medieval times, they have been added to jams and syrups. You may have encountered them in the form of fresh pansies sprinkled over salads, violets candied to garnish desserts, jasmine or rose petals dried to flavor tea, or as the deep-fried buds of squash blossoms in Mexican food. Flowers provide flavor and aroma—not to mention beauty—to ice creams, cakes, confectionery, salads, cocktails, and cordials.

Because of their fragility, fresh flower petals can be used only in cold or uncooked dishes like salads, or as a garnish for desserts. The actual flowers of herb plants—rosemary, dill, thyme, basil, anise hyssop—are delicately delicious, like an echo of the flavors of the herbs themselves. But even if you never cook with actual flowers in any form, you still frequently work with floral flavors. The aromatic molecules that make flowers taste and smell "floral," geraniol and linalool, are also found in many spices, herbs, and citruses. To understand how floral flavors function in cooking, it helps to look at their use in perfume. They work as smoothers, bridging the gaps between sharper ingredients like spices and making them rounder and softer.

FAMILY RESEMBLANCES LIE not just within each of these groups but beyond them, as with the "floral" spices, herbs, and citruses. Conversely, as with the genetic differences between human siblings, minute differences in chemical makeup can leave one family member worlds away from another, and even those that lie close together—lemon and lime, cinnamon and clove, mint and basil—differ from each other in significant ways. We group the flowers and the citruses each within a single family because overall they are more similar than they are different, with a relatively limited range to the family. Yet lime zest is green, round, and slightly sweet, while blood orange zest smells of raspberries, and grapefruit zest carries notes of pine—not to mention that no two instances of any of these will taste exactly the same. Substituting one for the other will create a completely different flavor.

The herbs and especially the spices cover such a range of flavors that we have grouped them into subfamilies according to their dominant notes, as a way of learning to appreciate their general effects. Yet even between varieties of a particular spice or herb, minute differences in chemistry have radically different results: the spicy notes of Thai basil lend themselves to a brightly acidic curry, while sweet basil is better suited to join quietly into a traditional Italian-style tomato sauce.

Family and subfamily descriptors help organize our thinking about flavor at a high level by creating broad categories of association. At a more detailed level—down in the weeds, so to speak—they allow us to concentrate on key distinctions. The point is to let you track which flavor characteristics make a given ingredient fit with its family members and which make it stand out from them, so that you can explore ways of making use of both their similarities and their differences. We have not attempted to be exhaustive, but have focused here on the most widely known and used flavor ingredients, and also a few that are lesser known but particularly useful. We hope that you will add many more to your repertoire, and will discover favorite varieties and sources for each. It helps to keep a family tree, but ultimately you have to get to know the individuals to discover which cousins you like best.

{ NATURE IS THE ORIGINAL FLAVORIST }

Nothing in nature is simple, and our compass can only hint at the complexity of flavor contained in even the most common herb. As we have seen, the flavor of any given ingredient is not that of a single molecule but a composite of the many such molecules it contains, some major, some minor, and some trace. Each unique combination creates a flavor as distinct as a snowflake, or a personality. That's why chemistry, crucial as it is, isn't a useful organizing principle for ingredients: a little difference is all the difference in the world. The same molecules show up in many different plants, but in different amounts, or in the company of an utterly different array of flavor-determining molecules. As the food science writer Harold McGee observes in a forthcoming book on the science of scent:

> Plants deposit countless different molecules in their wood and leaves and flowers and fruits, because they can. So the smell of any particular item is a composite of many different volatiles, of which perhaps a dozen or two predominate, and several different molecules can remind us of the same thing. And then on our end, we encounter many of the same molecules in different plants: so one molecule can remind us of several different things. In fact, that's part of the interest of paying close attention to flavor. When we do, we notice echoes and rhymes in very different things.

Oregano and thyme share not one but two major molecules, for example, carvacrol and thymol, but in different proportions and contexts, to markedly different effect. Basil and lavender, wildly divergent in taste and smell, share the same main aromatic molecule, linalool, but in different proportions and modified by the complex cocktail of the other aromatic molecules in the plants. Black pepper and lime share the molecule pinene—and there the resemblance ends.

You might say, in fact, that nature is the ultimate flavorist, and humans are but ploddingly emulating her as we cook. In this we are only sharing the worldview of the alchemists, who saw nature as "a guiding principle, whose ways were to be studied, understood and imitated in alchemical operations," notes Andrea de Pascalis in *Alchemy: The Golden Art*. "The alchemist does not invent, or create, he merely imitates and transforms through his Art that which nature produces." Nature isn't showing off for us, of course—organisms defend themselves with volatile compounds rather than create them for our delight—but nevertheless nature often creates flavor as we aspire to, by joining ingredients in ways that allow all their facets to interact as interestingly as possible, to create a distinctive taste. Like a pinch of spice thrown into a dish, sometimes the presence of a tiny amount of a certain molecule can make all the difference in how it interacts with other ingredients. In grapefruit, the molecule nootkatone is present in less than a half of a percent, but it plays a major role in making a grapefruit a grapefruit, and not an orange. So when we cook, we are not only combining, say, a dozen ingredients. We are really combining dozens, even hundreds, of ingredients within each of those ingredients. It could be tiny aspects of just a few of them that make the difference between good and out-of-this-world delicious.

According to Harold McGee, there are two particular chemical families that contribute the majority of flavor compounds to herbs and spices: terpenes and phenolics. In *On Food and Cooking*, he describes terpene compounds as "characteristic of the needles and bark of coniferous trees, of citrus fruits, and of flowers"; they "provide pine-like, citrusy, floral, leaf-like and 'fresh' notes to the overall flavor of many herbs and spices." These compounds, he notes, can have a generic quality. The phenolics, on the other hand, "are distinctive and define the flavor of such spices as cloves, cinnamon, anise, and vanilla, as well as the herbs thyme and oregano. The pungent compounds of chilis, black pepper, and ginger are also synthesized from a phenolic base."

Here are some of the common terpenes and phenolics that appear in essential oils, along with their dominant smell/flavor:

Terpenes

PINENE	PINE
LIMONENE	CITRUS
CITRAL	LEMON
GERANIOL	ROSE
LINALOOL	FLORAL
CINEOL	EUCALYPTUS
MENTHOL	PEPPERMINT
L-CARVONE	SPEARMINT
GERANYL ACETATE	ROSY FLORAL
CAMPHOR	MEDICINAL AND COOLING

Phenolics

EUGENOL	CLOVE
CINNAMALDEHYDE	CINNAMON
VANILLIN	VANILLA
THYMOL	THYME
CARVACROL	OREGANO
ESTRAGOLE	TARRAGON

Multiple compounds are present in a single ingredient, where nature does just what a perfumer aspires to—smoothing the sharp edges of citrusy citral in lemongrass with roselike geraniol, for example. At the same time, the same aromatic molecules are present in many herbs and spices that have completely different flavors. They contribute a facet to the overall aroma, creating echoes across many dissimilar herbs and spices. For example, eugenol is a large component of clove and gives it its dominant clovelike character, but it is also a facet of basil.

Here are some instances of the range of ingredients where key terpenes occur in addition to the ingredients or families whose dominant character they define:

CINEOL (EUCALYPTUS)—also in sage, cardamom, rosemary, galangal, lavender, basil, and oregano

CITRAL (CITRUS)—not only in citruses like orange, lemon, lime, and bergamot but also in lemongrass, ginger, rose, geranium, and basil

GERANIOL (ROSE)—also in bergamot, rose, thyme, orange flower, lemongrass, coriander seed, cardamom, and galangal

LIMONENE (CITRUS)—not only in lemon, lime, grapefruit, orange, and bergamot but also in dill weed, black pepper, thyme, spearmint, and rosemary

LINALOOL (FLORAL)—not only in orange flower and lavender but also in coriander seed, thyme, basil, bergamot, coriander leaf, jasmine, and sage

PINENE (PINE)—not only in resinous herbs like rosemary and sage but also in juniper berry, nutmeg, black pepper, coriander seed, cumin, lime, lemon, bergamot, and galangal

Aroma molecules, in other words, cross species boundaries like invasive weeds. Perfumers (and flavor scientists) actually have some of those constituent chemicals available to them as individual ingredients, in the form of natural "isolates" separated out from an essential oil by further processing, so they have a way to add a slice or dash of a single facet to a given composition by this means. The cook, however, picks up an entire shopping cart of ingredients each time she introduces a single one into a dish. Ultimately this is the great magic of cooking. In the dynamic of this complex marriage of facets lies the potential for great flavor.

But how to choose which of the intense array of aromatic ingredients to work with in a given dish? It is here that the distinguishing nuances—facets—become of primary importance. The soft, sweet facets of lemon will round off the rough edges of resinous rosemary, but lime, with its own more resinous, green, and angular qualities, may create too harsh a pairing. In nature as in cooking, proportion determines flavor: more geraniol in lemongrass makes it softer and rounder, while more citral makes it sharper and stronger. A rosemary that tastes sweeter and more approachable might be a particular variety with an extra dose of limonene. Growing intimately familiar with the facets of flavor ingredients—with the help of the compass and your own nose and tongue—will help guide you toward the best possible alchemy of ingredients.

{ SPICES }

When working with spices, it is important to keep in mind their generally strong inherent level of intensity, especially when combining them with other ingredients, including other spices. Within each subgroup here, we have ranked them in order of intensity from least to most intense.

Sweet Spices

Sweet spices share an overall characteristic of warmth and roundness, in both taste and aroma, that lends them to inclusion in sweet dishes. Their association with sweet foods can, in fact, make a dish seem "sweet" even apart from the inclusion of actual sweeteners. They can also balance savory, bitter, and pungent flavors.

• ANISE •

Anise, *Pimpinella anisum*, has been used at least since 1500 BC; the early Egyptians are known to have used it. The tiny fruits of the plant look like small gray seeds and are commonly referred to as such. They have a sweet licorice-candy character with a refreshing camphor facet. The warm and fruity taste is as delicate as the shape. Its character is determined primarily by anethole (anise), a phenol. It is best to buy the seeds whole and grind them right before use, as they quickly lose their flavor when ground.

• NUTMEG •

Nutmeg is the ovoid wrinkled dried kernel of the peachlike ripe fruit of *Myristica fragrans*. The name comes from the Latin *Nux moschata*, musk nut, so named for its

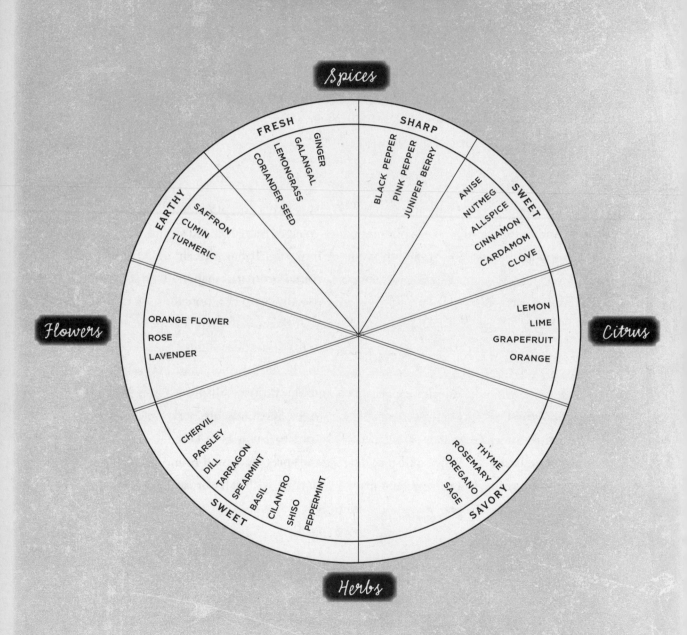

The Flavor Compass

Spices

Flowers

Citrus

Herbs

FRESH

GINGER
GALANGAL
LEMONGRASS
CORIANDER SEED

SHARP

BLACK PEPPER
PINK PEPPER
JUNIPER BERRY

EARTHY

SAFFRON
CUMIN
TURMERIC

SWEET

ANISE
NUTMEG
ALLSPICE
CINNAMON
CARDAMOM
CLOVE

ORANGE FLOWER
ROSE
LAVENDER

LEMON
LIME
GRAPEFRUIT
ORANGE

CHERVIL
PARSLEY
DILL
TARRAGON
SPEARMINT
BASIL
CILANTRO
SHISO
PEPPERMINT

THYME
ROSEMARY
OREGANO
SAGE

SWEET

SAVORY

musky sweetness. Nutmeg's character is deeply bittersweet, warm, woody, and balsamic, with clove facets. It has a rounded and powdery aromatic shape and medium to light intensity. Its sweet, woody warmth—it was once a component of Coca-Cola—is delicious in puddings and egg dishes, fruit tarts and pies, lemon desserts and chocolate. Nutmeg is also welcome in vegetable stews and pastas, especially as a delicate accent in dishes that contain cream and cheese. Since it loses flavor and fragrance quickly once ground, it should be bought whole, grated right before use, and added at the end of the cooking process, except in baking.

• ALLSPICE (PIMENTO BERRY) •

Allspice, or pimento berry, *Pimenta dioica*, tastes like a combination of cloves, cinnamon, and nutmeg—hence its common name. Warm, pungent, and sweet, its flavor is slightly peppery but not dry, with medium intensity. It has a fresh and clean top note, sweet balsamic and tealike facets, and a round aromatic shape. The dominant flavor molecule in allspice is eugenol, which is also the main component in clove, but pimento berry has sweeter floral facets that make it like a "softened" clove.

The berries are collected unripe (green), then dried in the sun until they look like very large black peppercorns. The flavor lies in the hull rather than in the seeds, which lose their aroma upon ripening. Aztecs and Mayans used allspice to flavor chocolate. It is a popular ingredient in Caribbean cuisines, most notably "jerk" treatments. In the Unites States it is most often used with tomato, such as in ketchup and some tomato-based barbecue sauces. Allspice's sweet and spicy facets and round shape allow it to pair well with sweet flavors, and provide contrast with salty or savory ones.

• CINNAMON •

Cinnamon, *Cinnamomum zeylanicum*, was once the king of spices and deserves reconsideration of its current status, which is largely limited to desserts. It was first employed in China around 2500 BC for medicinal purposes. Every part of the cinnamon tree has

ONIONS AND GARLIC

Although we haven't included them on the flavor compass, onions and garlic are the foundation of many dishes in almost every cuisine around the world—so familiar to most cooks that we chose not to focus on them on their own. By themselves, they can transform a dish, but their value lies also in their ability to heighten other flavors while remaining in the background.

Remember that the qualities of these trusty ingredients change with their state. Raw, they are sharp and aggressive. Marinated, they can be piquant but not abrasive. Charred, they are both smoky and sweet. Sautéed or sweated, they mellow, their flavor turning round and expansive. Either singly or in conjunction, they are the base on which many flavor combinations can be built. As in the Chicken Stewed with Saffron, Orange, and Tarragon (page 121), they can be a vehicle for uniting and amplifying other flavors. Learn to think about these versatile, indispensable ingredients not as an end in themselves but as the beginning of combinations with other, less used aromatics.

some use—bark, leaves, bud, roots, and flowers—but the most widely used is the bark. People often confuse cinnamon and cassia, which are closely related plants but have different flavor facets: cassia has coarser and thicker bark and shares with cinnamon a sweet spicy aroma, but is less lively, with a more pungent and harsh taste that has bitter facets. Cinnamon's character is sharply aromatic, warm, sweet, clovelike, and dry, with more delicate facets, but long cooking will bring forth bitter notes. Cinnamon possesses a note of clove from a small amount of eugenol, which is not present in cassia. Cinnamon is a wonderful addition to vinegars. The marriage of cinnamon and cooked tomatoes, common in Greek sauces, is surprisingly harmonious—actually downright fantastic—because of the way the sweet and spicy facets of the cinnamon meld into the sweet-sour fruitiness of tomato. If you want basil with that sauce, you'd typically reach for linalool-rich sweet basil, which would help smooth and sweeten the tomato sauce even further. But holy basil, with its eugenol-rich clove facet, would actually accent the spiciness of the cinnamon.

• CARDAMOM •

Cardamom, *Elettaria cardamomum*, has a warm and bittersweet, fresh, spicy, floral character, with facets of mint, eucalyptus, and black pepper. The major chemical components are cineol (camphor) and pinene (pine). Its aromatic shape is full-bodied and medium to high intensity.

Cardamom enhances both sweet and savory dishes. Toasting the whole pod intensifies its sweetness. Cardamom loses its flavor and fragrance quickly, so it is best to buy the whole pods and use them that way, or to crush the seeds in a spice grinder or with a mortar and pestle just before using. Cardamom is a beautiful flavor to add to coffee and to rice dishes, and it gives a green, lifting counterpoint to the sweetness of vegetables.

• CLOVE •

Cloves—from the Latin *clavus*, "nail" (because of their nail-like shape)—are the dried, unopened buds of the *Syzygium aromaticum* tree. Their dominant molecule is eugenol, and they pack a strong aromatic intensity. Clove aroma has a pointed shape and a character that is fruity, woody, burning sweet, assertive, and warm, with both peppery camphor facets and floral, carnationlike ones. You can tell if you have good-quality cloves by smelling them whole, then pressing them with your fingernail to see if the scent intensifies, which indicates that they are still rich in their essential oil.

Cloves can be used whole or pounded in a mortar and pestle to allow an even dilution of their inherent intensity. Used sparingly, they introduce welcome punctuation into an array of foods, including grains, legumes, meat stews, coffee, pickles, and baked goods.

Brisket with Tomato, Allspice, and Lime

SERVES 5–6

Spices love citrus, because citrus softens spices' sharp edges (like florals do). In return, spices make citrus more dynamic, angular, and energetic. Spices often have a citrus facet to them. Citrus and spices tend to lock to become one indivisible flavor.

Brisket is cut from the cow's hardworking breast muscles. It has lots of connective tissue and fat, and it rewards low, slow cooking with juicy, tender, and flavorful meat. Because it gives such bang for the buck, it is a favorite of many cultures, and comes with an array of cultural associations. It is browned and braised in a classic American pot roast, and boiled in the Italian bollito misto. It is cured and cooked into a St. Patrick's Day corned beef, and spiced and smoked for the pastrami at Jewish delis and the barbecue popular in its many variants all over the South.

The sweet, fatty nature of brisket makes it a wonderful partner for spices in all its manifestations, traditional and otherwise. It can be rubbed with them before slow cooking dry in an oven or on a smoker, but for most home cooks, braising is the easiest way to go, to break down the tough connective tissues and turn them rich and tender. Here allspice is the primary seasoning, but it needs balancing elements to make a satisfying savory dish. The bold spiciness of cayenne and black pepper offsets its soft sweetness, and tomato both gives body to the broth and creates a satisfying lock with the spices. Lime contributes a lifting top note, cutting through the fat and sweetness. Its angular acidity connects naturally with the mild acidity of the tomato, and it is sharp enough to stand up to the assertive spices.

3-POUND PIECE OF BRISKET

2 TABLESPOONS GROUND ALLSPICE

2 TEASPOONS FRESHLY GROUND BLACK PEPPER

1 TEASPOON CAYENNE

SALT

4 TABLESPOONS VEGETABLE OIL

2 MEDIUM ONIONS, CHOPPED

2 CARROTS, PEELED AND CHOPPED

4 TABLESPOONS TOMATO PASTE

4 CUPS WATER

FRESHLY GRATED LIME ZEST

FRESHLY SQUEEZED LIME JUICE

Preheat the oven to 250 degrees.

Trim the brisket of excess fat. In a small bowl, mix together the allspice, black pepper, and cayenne. Season the brisket well with salt, and then with the spice mixture. Let it stand for at least an hour or two in the refrigerator, preferably overnight.

In a Dutch oven or other large, deep, heavy pot, heat the vegetable oil over medium-high heat. Sear the brisket on both sides, being careful not to burn the spices. Remove the brisket to a plate, then add the onions and carrots and a pinch of salt to the pot. Cover and cook over low heat, stirring occasionally, until lightly browned and tender, about 10 minutes.

Dissolve the tomato paste in the water, then add that to the pot. Return the brisket to the pot, bring the liquid to a simmer, then cover the pot and place it in the oven. Bake until the brisket is tender, about 3 hours, turning the meat occasionally and adding a little water if necessary to keep the environment moist.

When the brisket is fork-tender, remove it to a cutting board. Bring the pan juices to a rapid boil and whisk to emulsify the fat into the sauce. Remove from the heat, add the lime zest and juice to the pan juices, and adjust the seasonings.

Slice the brisket and serve it over rice, with liberal amounts of sauce.

☞ The tomato paste in this recipe acts as a binder, thickening and uniting the fat from the beef and the cooking liquid. Broken-down or pureed vegetable fiber—tomato, onion, carrot—makes for a great emulsifier. The tomato also provides the sweet roundness that connects the spices and the beef.

Sharp Spices

Sharp spices—those with angular, even aggressive character—are essential to great flavor for their ability to provide shape and counterpoint, highlighting and lifting neighboring ingredients.

• BLACK PEPPER •

Black pepper, *Piper nigrum*, is the world's most ubiquitous spice. Its character is woody, clean, dry, and penetrating, with clove and lemon facets, and the presence of linalool gives it a floral facet as well. In fact, peppercorns are actually dried fruits, which, depending on the stage of ripeness at harvest time and processing, can be black, green, white, or red (pink peppercorns are not pepper at all; see below). Black and green peppercorns are the most aromatic, white the most pungent, green the mildest. Whole peppercorns will stay fresh for up to a year when stored in an airtight container, but they lose their aroma quickly once they are ground. Black pepper is welcome in a vast array of dishes, providing lift to a simple green salad and even brightening and focusing fruits and chocolate.

• PINK PEPPER •

Pink pepper, *Schinus molle*, is not a true pepper—its shape contributes to the confusion—but rather a member of the cashew family. Pink pepper's character is sweet, fruity, and red berryish, with clear pine and mild floral facets. Pink pepper is similar to juniper berry but with milder creamy and fruity pepper facets, similar to the top notes of black pepper. Pink pepper is best used in concert with other spices, contributing its floral and berry facets to the whole.

• JUNIPER BERRY •

Juniper berry, *Juniperus communis*, is the berrylike cone of the juniper bush. You can taste its characteristic flavor in gin: bittersweet with strong notes of pine and resin, refreshingly woody and astringent, with a lean and clean quality and warm balsamic notes.

Black Pepper – Crusted Steak with Cognac, Mustard, and Cream

SERVES 4

Black pepper's enduring popularity might be not only because of its astonishing versatility as a solo seasoning but also because of its role as a catalyst for and counterpoint to other ingredients. Here black pepper locks with mustard to create a counterpoint to the rich meld of beef, cream, and butter in one of the most time-honored dishes in the steak house canon, peppered steak with cognac and cream. Long relegated to establishments with frayed carpets and early-bird specials, the combination is worth dusting off to understand why it's so harmonious, and why it still works.

The bold, penetrating spice of black pepper, combined with the sharp mustard, serves as a counterbalance to a dish high in umami and fat even as they marry with the deep flavors of the browned meat. Bright herbs would have a similar effect, but would create too much distraction from the flavor of the meat. The meat is cooked in butter, which gently browns as the meat cooks, introducing additional aromatics. The trick to this is to keep the heat moderate, and to keep moving the steak around, so it doesn't develop burnt spots. After the steak is removed from the pan, cognac is added and the alcohol quickly burned off, before cream and mustard are introduced. The cream covers the spicy pepper and piquant mustard like a warm blanket, taking away their sting—though if you'd prefer a little more acidity, you can replace some, or all, of the cream with crème fraîche. The cognac provides a piercing aromatic note that lifts the whole.

2 POUNDS STEAK, PREFERABLY STRIP OR
    RIB EYE
SALT AND WHOLE BLACK PEPPER
2 TABLESPOONS BUTTER

2 TABLESPOONS COGNAC

1 CUP HEAVY CREAM

1 TABLESPOON DIJON MUSTARD

1 TABLESPOON WHOLE-GRAIN MUSTARD

Season the steak with a good amount of salt and lots of coarsely ground black pepper. Press the seasoning into the meat.

Melt 1 tablespoon of the butter in a nonreactive sauté pan over medium-high heat. Add the steak and cook, turning often to allow it to cook evenly, until it is well browned and at the desired degree of doneness. Be careful not to let the pepper burn.

Remove the steak to a plate and wipe out the pan to remove any excess salt. Add the cognac to the pan, then add the cream. Bring to a boil, then whisk in the mustards and the remaining butter. Adjust the seasonings. By the time the sauce is finished, the steak will have rested just long enough to be sliced and plated. The sauce should be spooned generously over the meat and any accompaniments, but don't overdo it. A bit of something green, like sautéed spinach or some raw leaves of watercress, provides a counterpoint to the heavy meat and sauce but not a distraction. A potato puree or gratin is a good complement as well, with the mild, accommodating flavor of the potatoes providing a pleasant accompaniment to both the steak and the sauce.

☞ This recipe uses both Dijon and whole-grain mustard, because they each bring different qualities. Dijon is milder, smooth, and subtle, while the bold whole-grain mustard gives the sauce a spark.

Fresh Spices

Some spices have a fresh, bright character that fades with long cooking. They have lower intensity than other spices, and they tend to be higher on the structural scale, on the borderline between middle and top notes. Some have a strong floral facet—for example, both lemongrass and coriander—and they can lock when combined with each other or with a floral ingredient like lavender or rose.

• GINGER •

Ginger, *Zingiber officinale*, comes from the Sanskrit *shringavera*, meaning "shaped like a deer's antlers." When fresh, its character is spicy, light, sweet and warm, delicate and slightly pointed. The presence of citral gives it a lemonlike facet. Fresh ginger is more aromatic than dried. The essential oil carries the distinctive aroma of ginger from the predominant volatile molecule zingiberene (around 70 percent when fresh), but ginger's characteristic heat or pungency comes from heavier molecules like gingerol, which are not volatile and therefore get left behind when the essential oil is distilled from the plant. Longer cooking increases fresh ginger's pungency but decreases its aroma. Most cuisines that have access to ginger use it fresh, though American cooks know it in both forms.

• GALANGAL •

Galangal root, *Alpinia galanga*, looks like ginger but tastes like a combination of cardamom, ginger, and saffron. It has a rich, spicy core, with bitter, lemony qualities and warm, woody pine and nutmeg facets. Its major component is eucalyptus (cineol) with a smaller amount of pine (pinene). Its complexity allows it to enhance a wide range of other flavors. The dried version is markedly different from the fresh: fresh is more like fir or pine needles, and dried is more like a cross between ginger and cinnamon.

• LEMONGRASS •

The character of lemongrass, *Cymbopogon citratus*, is tart, clean, and delicately lemony, balanced with grassy and herbaceous notes. Its dominant citrus character (70 percent citral) is tempered by facets of slight rosiness from some linalool, geranyl acetate, and geraniol. The flavor of lemongrass comes from the tender inner stalk, which can be accessed by smashing the stalk and infusing it in liquid. Soft and well rounded, it enlivens other ingredients without dominating.

• CORIANDER SEED •

The name coriander, *Coriandrum sativum*, comes from the Greek word *koris*, meaning "bedbug," reportedly because of the "buggy" odor of the unripe seeds. Coriander's character is sweet, floral, woody, and peppery. It has floral facets from a large percentage of linalool along with geraniol and geranyl acetate (rose). The seeds do not share the distinctive love-it-or-hate-it soapy, green smell of the leaves, otherwise known as the herb cilantro. The seeds develop their own, more intense aroma when they are dry roasted.

Pork Shoulder Simmered with Coconut Milk, Ginger, and Lemongrass

SERVES 6

The flavors in this recipe are typical of Southeast Asia, but sometimes "typical" happens for a reason. This combination of flavors is exceptionally delicious. Coconut milk cooked into the broth creates a sweetly floral creaminess, which melds with the pork to create a flat, rich, unified middle note. Ginger's bright, lemony-spicy facets work well with the citrusy lemongrass to provide complexity, but can't fully lift the combination. Chili peppers add heat to the base and serve as a counterpoint to the sweetness. The aromatics are added incrementally, to preserve the freshness they contribute. But it's the top note of lime, with its sharp, green acidity, that helps define the broth and give it life. The basil is sweet and bright, and its aroma as it melts into the warm broth gives the dish its final lift and enticing first impression.

3 POUNDS PORK SHOULDER, CUT INTO 1-INCH CUBES

SALT AND FRESHLY GROUND BLACK PEPPER

VEGETABLE OIL

2 MEDIUM ONIONS, SLICED

1 RED BIRD'S EYE CHILI PEPPER, CHOPPED

2 CANS (28 OUNCES) COCONUT MILK

1 CUP WATER

1 STALK LEMONGRASS, SMASHED WITH THE
   BACK OF A KNIFE

1 TABLESPOON MINCED GINGER

FRESHLY GRATED LIME ZEST

FRESHLY SQUEEZED LIME JUICE

THAI BASIL

Season the pork with salt and black pepper. Place a Dutch oven or other large, heavy pot over high heat and film it with oil. Add the meat and sear it in batches, being careful not to crowd the pan. Remove the meat as it is browned, then reduce the heat to low and add the onions and chili pepper. Cook, stirring often, until the onions and pepper are softened, about 5 minutes.

Return the pork to the pot and add the coconut milk and water. Cover and simmer for 45 minutes, then add the lemongrass. Simmer for another 30 minutes and then add the ginger. Simmer until the meat is tender, an additional 15 to 30 minutes.

Remove the pan from the heat and discard the lemongrass. Add lime zest and juice to taste and adjust the salt. Just before serving, stir in a generous handful or two of torn basil leaves.

☞ Sweet basil will work here, but try using Thai basil if you can find it. Its sharp, anisic qualities provide a more energetic counterpoint.

Earthy Spices

Spices that tend to have a dampening, flattening character are those we think of as earthy. Their musky, sweaty, flat aromas and tastes don't lift, but rather sink down and lock with deeper, richer, meatier elements. Earthy spices work best in concert with bright flavors such as citrus and sweet spices.

• SAFFRON •

Saffron, whose name is derived from the Arabic *za' farān*, which means "yellow," is the stigma—the female sexual organ—of *Crocus sativus*. It is the most expensive spice in the world because it is so labor-intensive to harvest. Each plant produces several flowers per year, and each flower yields only three reddish-purple threadlike strands; these stigmas must be collected by hand and carefully dried to produce the spice. Its character is rich, delicate, and lingering, with musky and bitter facets, a warm floral bouquet, and earthy, honeylike notes of deep red fruits and oranges. Saffron works best infused in water or other liquids, to bring out the full spectrum of its fragrance and flavor before it is added to a dish. It imbues the other ingredients with a magical flavor and aroma as well as a gorgeous golden-orange hue.

• CUMIN •

Cumin, *Cuminum cyminum*, is a small, gray-green seed. Its name comes from the Sanskrit *sughandan*, meaning "good smelling." Powerfully dense and heavy, cumin has high flavor intensity, with complex facets that can lead it in many directions. It is warm and earthy, rich and fatty, but also bitter, with a persistent acrid, sweaty, gamey pungency. Toasting it dry or roasting it in oil renders it more nutty and earthy, with a richer, more mellow flavor and a more intense aroma.

• TURMERIC •

Turmeric, *Curcuma longa*, is a rhizome usually sold dried, but it can be used fresh as well. Fresh turmeric is musky, earthy, and more pungent than the dried version, with pepper and ginger facets. Sometimes the dried form reveals an unpleasant medicinal aroma. Fresh turmeric helps to connect and smooth other spices. Like saffron, it imparts an intense color—turmeric is more yellow than orange.

Chicken Stewed with Saffron, Orange, and Tarragon

SERVES 4-6

This recipe is a template for layering flavors: a spice, a citrus, and an herb. The base for the dish is the chicken and onion. The middle is the spices. The top note is the citrus and herbs. Here, orange teases out the latent orange notes in the saffron, and is sweet and mild enough not to overpower the elegant aroma of the spice. The tarragon introduces a counterpoint of green licorice, again without overpowering. There is a lot of sweetness in the dish, so a little bit of chili and vinegar balance the flavors and keep them savory. Variations could include earthy cumin combined with the clean, bright lift of lime and the funky, almost animal notes of cilantro. Or to go in a sweeter, more modulated direction, try it with nutmeg, lemon, and basil. When you contemplate citrus options, however, steer clear of grapefruit in this treatment; its bitter, piney, resinous flavors work better in fresh compositions or as a finishing note.

2 POUNDS CHICKEN LEGS

SALT AND FRESHLY GROUND BLACK PEPPER

3 TABLESPOONS BUTTER OR NEUTRAL-TASTING OIL

2 MEDIUM ONIONS, SLICED

1 TEASPOON SAFFRON THREADS

½ TEASPOON CHILI FLAKES

2 CUPS WATER

¾ CUP FRESHLY SQUEEZED ORANGE JUICE

4 TABLESPOONS CHOPPED FRESH TARRAGON

FRESHLY GRATED ZEST OF 2 ORANGES

2 TABLESPOONS RICE WINE VINEGAR

Preheat the oven to 250 degrees.

Season the chicken legs generously with salt and black pepper. Heat the butter or oil over medium-high heat in a braising pan, add the legs, and brown them all over. Add the onions and a little more salt. Stir in the saffron threads and chili flakes, put the chicken legs on top, and add the water and orange juice. Bring the liquid to a boil, then cover the pan and put it in the oven.

When the chicken is tender, after about 1½ hours, remove it from the oven. The saffron will have turned the chicken a beautiful yellow hue. Stir in the chopped tarragon and orange zest, and season the sauce to taste with the rice wine vinegar and salt. Serve over rice.

☞ The onions give the sauce body, but there will be a lot of free-floating fat from the chicken and the searing. If you prefer a smooth sauce, before you add the tarragon, transfer the contents of the pan to a blender or food processor, process till smooth, then return the mixture to the pan and sprinkle on the herb. As with the tomato paste in the Brisket with Tomato, Allspice, and Lime (page 110), the onions emulsify the fat and liquid.

{ HERBS }

Sweet Herbs

By "sweet" herbs we mean the "toppiest" of top notes, the herbs that call out a bright greeting that introduces every flavor that follows. Their function is to lift and vivify the way a dish opens—how it initially presents itself to mouth and nose. In this role, they get along easily with most other flavors. The one difficult sweet spice is peppermint, because of the dominant—often overbearing—presence of menthol.

• CHERVIL •

Chervil, *Anthriscus cerefolium*, is a delicate, fragile herb with a light, sweet character and an aromatic, aniselike flavor. It has facets of caraway, pepper, and parsley—and can in fact be used as a more anisic version of parsley. It is used much in the same way as parsley and cilantro, but it is considerably less intense even than parsley—somewhere between herb and salad green. As well as adding a subtle touch of green to a dish, it has a delicate anise flavor that marries well with salads and light vegetables. It needs to be added at the end of cooking or, preferably, used raw.

• PARSLEY •

Don't bother with curly parsley, or dried parsley in any form; they have almost no flavor. When a recipe (or your own impulse) calls for parsley, use flat-leaf parsley, *Petroselinum crispum*, whose leaves have a somewhat more angular shape than those of cilantro. Flat-leaf parsley is tangy, grassy, green, and herbaceous, with a minor lemon facet. Because of its fresh green aroma, it has long been used as a breath freshener.

Parsley is widely popular because of its ability to freshen a dish without imposing a strong will. With its low flavor intensity, it can lock with other bland herbs like chervil or even help hold assertive herbs like cilantro in check. Learn to employ it as more than a default choice, to support other flavors or, used in profusion, to introduce a strong grassy note.

• DILL •

Fresh dill, *Anethum graveolens*, has a clean, mild, and sweet-spicy character, with anise and lemon facets. The citrus facets come from the significant presence of limonene. Dill weed spreads out and diffuses in a flavor, as feathery as its physical appearance. Because of its clichéd association with pickles, it requires the cook to approach it with a creative mind. Its soft, bright greenness gives a lift to delicate ingredients like chicken, carrots, and salads without overpowering them.

• TARRAGON •

Tarragon, *Artemisia dracunculus*, has a cool, licorice-like aroma. Its character is green, sweet, and slightly like celery, with anise/basil facets that it shares with basil. It partners not only with fennel and aniseed but with almost all other flavors as well. Tarragon has long, narrow gray-green leaves. Fresh tarragon is stronger than dried, which is delicate and more haylike. Fresh tarragon should be added near the end of cooking to keep its flavor.

• SPEARMINT •

Like peppermint, spearmint (*Mentha spicata*) has a sharply pointed shape and an almost biting quality, but the much lower presence of menthol makes it more amenable to use in composing flavor. Spearmint also has a fruity facet of mangoes and pears from a tiny amount of the aroma molecule myrcene. This profile gives it a warm, refreshing character, with sweet, lemony aspects and a much broader range of applications in both sweet and savory dishes. Heat dissipates its flavor, so it is best added at the end of cooking or used raw.

• BASIL •

The overall character of basil, *Ocimum basilicum*, is green, sweet-spicy, and peppery, with facets of cooling clove and anise. There are more than sixty species of basil, each with its distinct chemical makeup and resulting facets. Sweet basil, the most common variety, is floral and delicate because of the major presence of linalool. Holy basil contains more eugenol (clove), which makes it spicier, and Thai basil has more methyl chavicol, which gives it more of a licorice facet. Basil leaves have dotlike glands that contain the essential oil, but the oil in the flowers is the best quality: if you grow the plant, let it flower and use the blossoms! The different varieties of basil vary enormously, and in many instances one cannot be substituted for another. Delicate sweet basil, for example, does not substitute well for spiky Thai basil in a curry. Lemon basil is terrific in a dish that also includes lemon juice, as the lemon facets will lock together. The delicate flavor and aroma of all forms of basil decrease with cooking, so it is usually added at the end of cooking.

• CILANTRO •

Cilantro, *Coriandrum sativum*, is the broad leaf of the coriander plant. The leaves look much like flat-leaf parsley, but are rounder and with more delicate teeth along the edges. Cilantro's character is complex: soapy, piney, roasted anise with lemon and pepper facets. It needs to be added fresh at the end of cooking to preserve its pungency and bright color. While the seed is rich in the floral aroma molecule linalool, the leaf does not have that floral facet. Cilantro is a polarizer: people either love it or hate it. Those who hate it say that it tastes like soap to them—a perception that might be hardwired.

• SHISO •

Shiso, also known as perilla, *Perilla frutescens*, is a relative of mint and basil; it is a garnish familiar from many plates of sushi. It has a strongly aromatic, warm, spicy, sharp, citrusy, angular character, with facets of basil, anise, and cinnamon. Because it has so many different aromatic facets, it can be pushed in many directions, depending on what you choose to combine it with. With sharp herbs, its pointed flavor lends highlights, but with a round floral like rose, it recedes and makes it spicier. It goes well with the sweet/metallic flavor of turnips, folded into a sauté, or sprinkled on a soup. It works wonders with tomato and watermelon, at once lifting with its floral herbaceousness and punctuating with its spiciness. Or try it with peach, perhaps to accent a compote.

• PEPPERMINT •

Peppermint, *Mentha piperita*, is so powerful and distinct—and so linked in our common experience to mints and candy canes—that it is hard to find a place for it in savory cooking. Peppermint's flavor profile is more than 50 percent menthol, whereas spearmint's percentage is very low. Its fiery bite and pure, refreshing aroma are rousing, fresh, and cooling, with tangy peppery facets. A tiny amount of jasmone (a naturally occurring jasminelike molecule in some peppermints) helps lend a sweet facet to soften the peppermint's harshness. Peppermint should be used judiciously and sparingly, and added at the end of cooking, as heat dissipates its flavor.

Roasted Beets with Goat Cheese, Rye Bread, and Dill

SERVES 4

Sweet herbs are connectors, bridging disparate flavors the way that citrus and florals do. The sweetness of these herbs can smooth bitter edges or, as in this recipe, leaven earthy flavors. Here dill makes a traditional pairing with beets, which, although they can partner well with many different foods, can be finicky. Sometimes they are as sweet as candy, other times earthy and bitter, and there is no way to tell which you've gotten when you're buying them, even if they're field-fresh and of pedigreed provenance. Before you incorporate them into this recipe, get them the way you like them: roast them, then peel and slice them and season them with sugar, a little salt, and just a touch of vinegar until they suit your preferences—lean and earthy or on the sweet side.

Peppery arugula punctuates the earthy base note of the beets, as does the spiciness of the thinly sliced radishes. Tangy, rich fresh goat cheese binds everything together, its creamy mouthfeel complementing the dense, sturdy texture of the beets, with the sweet dairy and sweet vegetable wrapping around each other in a very harmonious way. There's a reason this combination is so popular! From there the salad needs only a little texture—rye croutons, with notes of caraway—and a top note: sprigs of dill, which carry a citrus note and a slightly bitter edge. It's that relationship with caraway, one of the main seasonings in traditional rye bread, that makes dill a harmonious partner to rye. If you opt for mint or tarragon instead of dill here, choose a more neutral kind of bread for the croutons.

1 BUNCH (4–5 MEDIUM) RED BEETS

VEGETABLE OIL

SALT

DAY-OLD RYE BREAD, CUT OR TORN INTO ½-INCH CUBES, CRUST REMOVED,
    TO MAKE ABOUT 1 CUP

SUGAR AND CHAMPAGNE VINEGAR AS NEEDED TO SEASON BEETS

1 HEAD ENDIVE, SLICED

2 TABLESPOONS FRESHLY SQUEEZED LEMON JUICE

2 TABLESPOONS OLIVE OIL

FRESHLY GROUND BLACK PEPPER

2 OUNCES FRESH GOAT CHEESE

6 RADISHES, SHAVED THIN

FRESH DILL

Preheat the oven to 400 degrees. Scrub and dry the beets, then toss them with just enough vegetable oil to coat them; add a little salt. Place them in a roasting pan with a lid, add water to a depth of a quarter of an inch, and roast, covered, until the beets are tender, about 40 minutes, shaking the pan every once in a while and checking to see if you need to add water. Make sure the beets are completely tender. If they are even a little bit crunchy, their sweetness won't be fully revealed. When they are done, let them cool to room temperature.

Turn the oven down to 300 degrees. Spread the bread cubes on a baking sheet and put them in the oven. Stir them frequently and remove as soon as they are crisp and browned, about 10 to 15 minutes. Remove them from the oven and let them cool.

When the beets are cool, slide off their skins. Cut them into bite-size chunks and season them with salt, sugar, and vinegar to taste. Then toss them with the endive, lemon juice, and olive oil, and season with salt and black pepper, adjusting quantities to taste. The salad should be brightly seasoned, with a good depth of flavor and balanced acidity.

Put the beets in bowls, and dot with pieces of goat cheese, shaved radish, and the croutons. Sprinkle dill generously over the top.

☞ Sometimes a dish requires more than one kind of acidity. Here the champagne vinegar enlivens the beets and balances the sweet-earthy notes. Then, for the salad itself, the beets are tossed with endive and lemon juice for a more gentle sweet-sour flavor.

Savory Herbs

Some herbs have a strong resinous quality, almost like sap. Resins like frankincense and myrrh have historically been important perfume ingredients—burned as incense as far back as biblical times. Resinous herbs are not quite that strong, but their dense, powerful flavors and aromas usually need to be cooked to mellow them out. We call these herbs savory, and think of them as the closest of all the herbs to spices. Like spices, given their high intensity level, they should be used in smaller amounts. And like spices, they retain their essential oils well, and are sometimes as good in dried form as in fresh—or even better. This may be because drying collapses the cellular structures in the plant tissue, releasing the essential oil within the plant.

• THYME •

The distinctive character of thyme, *Thymus vulgaris*, is determined primarily by the presence of about three times as much thymol as carvacrol. Thyme is sweet, warm, herbaceous, dry-woody, and biting but not bitter, with facets of clove and camphor. A particularly beautiful variety is lemon thyme, which, as its name suggests, is rich in

THE ART OF FLAVOR

lemon facets (citral) and also rose (geraniol), compounds that smooth the bitter herbal facets into something brighter and softer. Unlike some of the other resinous herbs, it loses its aroma quickly with heat.

• ROSEMARY •

Rosemary, *Rosmarinus officinalis*, is bitter and complex, filled with major molecules of camphor (borneol), eucalyptus (cineol), and pine (pinene). The complexity comes from these molecules, along with its lesser facets of pepper, clove, and sage. Sometimes it also has a bitter, woody aftertaste. Rosemary is one of the sturdiest herbs; it doesn't lose its flavor even when cooked for a long time, though cooking tames its pine notes.

• OREGANO •

Oregano, *Origanum vulgare*, is one of those rare herbs that is more potent dried than fresh. Oregano's character is sharp, bitter, herbaceous, and warm. As noted, its character drives primarily from thymol and carvacrol, just like thyme, but in differing proportions and with different aromatic bit players. To a lesser degree, the proportions, and therefore the flavor, mild or strong, vary within the species itself, depending on the origin and growing conditions of the oregano—your mileage may vary.

• SAGE •

Sage, *Salvia officinalis*, has a penetrating odor and slightly bitter taste that is initially cooling, but then turns warm and spicy. Sage's character is herbaceous, bitter, strong, musky, and balsamic, with a sweet eucalyptus facet and a bittersweet aftertaste. Its high composition of camphor makes it astringent in the mouth. Dried sage has a stronger flavor than fresh.

Brussels Sprouts Fried with Garlic and Sage

SERVES 6–8 AS A SIDE DISH

Savory herbs change when cooked, becoming softer and more mellow. Cooking diminishes the bitterness and allows them to connect more easily with sweet and flat ingredients, like meat and vegetables. Savory herbs can also be used as a counterpoint, like pepper, to season sweet elements, as sage does with butternut squash.

Although their popularity has climbed in recent years, brussels sprouts remain sadly underappreciated. They are a cruciferous vegetable, in the same family as cabbage and cauliflower. One of the less charming family characteristics is a tendency to produce a gassy, sulfurous aroma when boiled. Traumatic reactions to brussels sprouts usually stem from encounters with the boiled variety.

Brussels, like cabbages, are terrific shaved raw and tossed with vinegar and olive oil as a salad. And, like cabbages, they are great fried, sautéed, or roasted over high heat, which brings out a savory depth, intensifies their sweetness, and mutes their sulfurous qualities.

The dense texture of raw young brussels sprouts has something in common with that of raw artichokes, which was the inspiration for this recipe, based on a classic artichoke preparation from the Jewish quarter in Rome.

The brussels sprouts are browned in lots of olive oil flavored with whole cloves of garlic and sage leaves, then roasted in the oven until tender. The garlic is lightly crushed with a pan or the back of a knife to allow it to release its flavors quickly but not burn in the hot oil, and fried just long enough to release its aroma. The sage is fried in the oil first, too, muting its intensity and softening its bitter, resinous edge but leaving its hauntingly complex aroma full of menthol and briars. The sage locks with the browning garlic to create a deep herbal umami flavor that infuses the oil. You may have to do this in batches if you're making a lot (and you should make a lot; these are addictive). The recipe is enough to fit one 10-inch sauté pan.

2 POUNDS BRUSSELS SPROUTS

1–2 CUPS OLIVE OIL

30 CLOVES GARLIC, LIGHTLY CRUSHED

2 BUNCHES FRESH SAGE LEAVES

SALT AND FRESHLY GROUND BLACK PEPPER

Rinse and trim the brussels sprouts. Quarter them if large, halve them if smaller. Heat the oil until almost smoking and add the garlic. Shake the pan so the garlic colors but does not burn. After several seconds—as soon as you can smell the garlic—add the sage and shake the pan. The sage will immediately seize and curl in the heat and become very aromatic. Add the brussels sprouts and season them well with salt. Lower the heat to medium high and cook the brussels sprouts for several minutes, stirring frequently, until they are irregularly browned and delicious. If they are not quite tender, put them in a hot (400-degree) oven for a few minutes. Finish with black pepper and additional salt, if needed.

☞ This recipe is based on a traditional Italian preparation for cooking artichokes. Seeing the similar properties in disparate ingredients—in this case, the similar size and density of baby artichokes and brussels sprouts—is one way to imagine new flavor connections.

Citruses

• LEMON •

Lemon is light and tart, with a thin, sharp shape. The juice is both fruity and refreshingly sour, and is more commonly used than the peel, but as with other citrus, it's the peel that contains the highest concentration of essential oils and delivers the most intense kick. When grating the peel, be careful to use the brightly colored outer layer and not the spongy white inner pith, which is bitter rather than pleasantly sour. Lemon is not especially dimensional in flavor, but it is almost universally welcome for its ability to bring lift, cut heaviness, and temper sweetness, as we'll explore in chapter 8.

• LIME •

Lime is more angular, more aromatic, and more full-bodied in character than lemon. It has piercing and sharp acidity, with high-toned green and spicy facets. Of all the citruses, lime has the most heightened ability to cut through rich flavors. Substituting lemon for lime is inevitably disappointing.

• GRAPEFRUIT •

Grapefruit can be eaten out of hand, but it is still highly acidic. As noted, it has distinctive flavor and aromatic facets that set it, in all its varieties, a bit apart from other citrus. As described by Harold McGee, it has "an especially complex aroma, which includes meaty and musky sulfur compounds." Grapefruit oil, best cold-pressed from the peel, has a pear-mango facet, thanks to the aroma molecule myrcene. The most significant aroma molecule in grapefruit oil, however, is nootkatone, which gives it a woody, cedarlike note even though it constitutes only 0.2 percent of the flavor profile—a solid instance of small aromatic facets making all the difference. Because of the powerful taste of the peel, cooks often use the juice alone.

• ORANGE •

Oranges are used for their peel and their juice, though the juice is much less intense than the bitter peel. The main varieties of orange are sweet, bitter, and blood. Sweet orange is heavy, rich, sweet yet fresh, and powerful. Bitter orange, more often found at farmers' markets than in grocery stores, is much more acidic, fresh yet bitter, slightly floral and not sweet. Blood orange has ruby-streaked flesh and raspberry notes. The character of each is dominated by limonene (citrus), but sweet orange has facets of lemon (citral), floral (linalool), and pine (pinene), whereas bitter orange has a larger floral (linalool) facet, and blood orange has pine (pinene) and berry facets. This means that you can choose among not only citruses but also among oranges, depending on whether you want to slant your flavor toward sweet and complex (sweet orange), floral and dry (bitter orange), or a rich berry taste (blood orange).

Asparagus with Citrus Sauce

SERVES 2-4

All the citruses combine well with one another, with the exception of grapefruit. Grape-fruit likes only lemon, and even then only when the proportion of lemon is so low as to be indiscernible. In that role, the lemon works under the radar, making the grapefruit flavor sharper and clearer.

The other citruses can be combined in various ways, with a range of effects. In this sauce they combine to make a bright, beguiling sauce for asparagus. Because of the way their flavors lock, it's impossible to tell where one ends and the next begins. Butter is a medium to smooth the sauce and unite it with the asparagus.

1 BUNCH ASPARAGUS

1 TABLESPOON BUTTER

SALT

1 TABLESPOON WATER

2 TABLESPOONS FRESHLY SQUEEZED ORANGE JUICE

1 TABLESPOON FRESHLY SQUEEZED LEMON JUICE

1 TABLESPOON FRESHLY SQUEEZED LIME JUICE

Trim the asparagus. Melt the butter in a sauté pan over medium-high heat. Add the asparagus and a little salt, tossing to coat the asparagus. Cook for 1 minute, then add the water, cover, and cook until almost tender, about 3 to 6 minutes, depending on the thickness of the asparagus. Add the citrus juices and cook just until the asparagus is tender. Plate the asparagus and top with the sauce.

This sauce goes very well with fish. In fact, adding a piece of steamed white low-fat fish like halibut or snapper turns this dish into an entrée.

Flowers

Floral flavors are wonderful as smoothers and connectors, rounding sharp edges and softening harsh flavors. Mostly home cooks encounter them as facets of nonfloral ingredients, but they make wonderful additions to the flavor palette in their own right. As noted, the flowers of herbs are terrific ingredients if you grow your own. Rose and orange flower waters are widely available, and even dried and fresh edible flowers can be found in spice shops and ethnic groceries and online.

• ORANGE FLOWER •

In cooking, orange flower always refers to the flowers of the bitter orange tree and not the sweet orange tree. It is most easily available to home cooks in the form of orange flower water, a hydrosol that is the product of distillation that yields the essential oil neroli; it is used in pastries, confectionery, and Middle Eastern cooking. Orange flower is flowery, rich, and warm, but also delicate and slightly dirty. Its taste is bitter but its aroma is layered and exotic. It can be used sparingly in fruit salads, stews, candies, and puddings, and as a counterbalance to clove, ginger, and cinnamon.

• ROSE •

The heavy, sweet, soft, round, deeply floral, and sinuous character of rose make it a wonderful alchemist in a range of dishes, both sweet and savory. Like orange flower water, rosewater—a by-product of the distillation of roses for rose essential oil, or rose attar—is fairly widely available. Edible dried rose petals are also relatively easy to find in ethnic and spice shops and online. In addition to being a common flavoring in desserts and preserves around the world, it is a hallmark of Indian, Eastern, Turkish, North African, and Southeast Asian cuisines, where it is used in curries, rice dishes, spice mixtures like Moroccan *ras-el-hanout*, in pork dishes in Northern China, and even to flavor tea, chocolate, coffee, and wine. Oil of rose was a minor but significant contributor to early recipes for ginger ale, to balance the ginger. As in perfume, rose in the kitchen is astonishingly variable and layered. It is a great smoother and blender; a tiny amount takes the rough edges off other ingredients, knitting them together. It has a soft sweetness that allows it to merge with spices.

• LAVENDER •

Lavender, which can be used fresh or dried, and for both its flowers and its leaves, walks the line between flower and herb. Its character is sweet, floral, herbaceous, and refreshing, with balsamic-woody facets, a slight bitterness, and a spiciness, thanks to the presence of camphor. The leaves are more bitter than the flowers and they are a bit too harsh for many preparations. They are best in infusions like meat sauces, added at the end of cooking and removed once their flavor becomes discernible. Lavender has a high odor intensity and can be bitter, so it must be used with a light hand, but it is wonderful in both sweet dishes and savories, providing a refreshing counterpoint to creamy ice cream or cakes or as a garnish.

Many of the recipes in Robert May's 1660 *The Accomplisht Cook* call for rosewater, along with saffron, orris, and cinnamon. The book is a reminder of a time when flowers were considered essential not only for fragrance but also for flavor, and it contains a number of recipes that feature them, including these:

TO PICKLE ANY KIND OF FLOWERS

Put them into a gally-pot or double glass, with as much sugar as they weigh, fill them up with wine vinegar; to a pint of vinegar a pound of sugar, and a pound of flowers; so keep them for sallets or boild meats in a double glass covered over with a blade and leather.

TO CANDY FLOWERS FOR SALLETS, AS VIOLETS, COWSLIPS, CLOVE-GILLIFLOWERS, ROSES, PRIMROSES, BORRAGE, BUGLOSS, &C.

Take weight for weight of sugar candy, or double refined sugar, being beaten fine, searsed, and put in a silver dish with rose-water, set them over a charecoal fire, and stir them with a silver spoon till they be candied, or boil them in a Candy sirrup height in a dish or skillet, keep them in a dry place for your use, and when you use them for sallets, put a little wine-vinegar to them, and dish them.

TO MAKE A PASTE OF VIOLETS, COWSLIPS, BURRAGE, BUGLOSS, ROSEMARY FLOWERS, &C.

Take any of these flowers, pick the best of them, and stamp them in a stone mortar, then take double refined sugar, and boil it to a candy height with as

much rosewater as will melt it, stir it continually in the boiling, and being
boiled thick, cast it into lumps upon a pye plate, when it is cold, box them,
and keep them all the year in a stove.

A RECIPE FOR ROSE VINEGAR

Keep Roses dried, or dried Elder flowers, put them into several double
glasses or stone bottles, write upon them, and set them in the sun, by the
fire, or in a warm oven; when the vinegar is out, put in more flowers, put out
the old, and fill them up with the vinegar again.

Strawberry and Rose Drink

SERVES 4-6

Few ingredients conjure the lush, ripe smells of midsummer as successfully as strawberries and roses. But even though the feelings of pleasure that they evoke are simple, their constituent aromatic qualities are very complex. Strawberries vary according to place and type, and can have a round, full-bodied sweetness with a trace of underlying acidity and a raft of floral notes floating on top with echoes of cooked apricots, citrus, flowering thyme, and even roses.

This recipe relies, of course, on good strawberries. They should be ripe and very aromatic. The dried rose petals produce a flavorful infusion that locks with the floral notes of the strawberry, forming a perfumey taste that is neither one nor the other.

Something, however, is needed to balance the sweetness of the combination and also to round out the bitter edge of the rose-petal infusion. Lemon, which shares with rose the aroma molecule geraniol, does great things to both the strawberries and the rose, lifting the combination and giving it life without imposing itself. The drink needs the push-pull of sweet and sour, or it will taste flat.

1 POUND STRAWBERRIES

2 TABLESPOONS ROSEWATER

2–3 CUPS WATER

¼ CUP SUGAR

3–4 TABLESPOONS FRESHLY SQUEEZED LEMON JUICE

Wash and hull the strawberries. Puree them with the other ingredients, starting with the lesser amount of water, and taste. Adjust if needed. Depending on the flavor of the fruit, the amounts of the other ingredients will vary considerably. If the strawberries are densely flavored and intensely sweet, the mixture might need more water (to dilute enough to make a pleasant drink), more rosewater and lemon juice (to balance the intensity and sweetness of the strawberries), and less sugar. For off-season or milder strawberries, the opposite might be true. The drink should be bright and fresh, like good lemonade.

Pass the puree through a strainer to remove the seeds. Chill completely and serve over ice.

☞ If you can find them, use dried rose petals instead of rosewater: Heat the water to a boil and pour it over a few tablespoons of petals to steep like a tea. Strain before adding.

This chapter has introduced you to some of the most potent tools for flavor in the cook's arsenal. As you get more familiar and comfortable working with them, you'll learn how they not only combine well, but also occasionally achieve a nirvana effect that gives a dish an unquantifiable, more-than-the-sum-of-its-parts transcendence. It's the specific mechanisms that get to that sweet spot (in all senses) that we'll explore in the next chapter.

six

ALCHEMY: LOCKING AND BURYING

The chemist thinks of matter in terms of its invariant atomic or molecular constitution. Thus water is H$_2$O, and salt is sodium chloride. For the alchemist, by contrast, a material is known not by what it is but by what it does, *specifically when mixed with other materials, treated in particular ways, or placed in particular situations.*

TIMOTHY INGOLD, *MAKING: ANTHROPOLOGY, ARCHAEOLOGY, ART AND ARCHITECTURE*

{ LOCKING }

What is this magical phenomenon of "locking" that we keep invoking throughout the book, and why is it such an important concept in creating flavor? Locking is the concept Mandy came up with to describe what happens when ingredients combine with impact that seems to be more than the sum of their individual characters. Usually this is because the facets of the individual ingredients are bonding

A recipe from *The Sultan's Book of Delights* is a veritable virtuoso display of mastery over locking ingredients, not to mention imaginative presentation:

Another method for cooking sour-oranges: take a really fresh orange and remove the top of the skin that is nearest the stalk. Cut a small ring from it, make a hole in that side of the orange and remove the pips, pith and flesh. Stuff it with the following ingredients: take two tulchas of well-pounded cardamoms, then grind half a miisa each of spikenard, wild spikenard, tagri, barmii, dried peel of sweet-orange, zerumbet, cloves and mutha beans and one miisa of saffron. Sift them in a sifting cloth and add one tulcha of ground musk, one diram of white ambergris, and two tulchas of camphor. Grind them all thoroughly with one tulcha of finely ground sesame seed. Collect all these ingredients in the sifting cloth and moisten it with rosewater. Perfume sandal with jasmine and put one miisa into the above ingredients. Stuff the orange with all of it, place the ring of orange, that had been cut off, on top and seal the top of the orange with dough. Rub the top of the dough with flowers. Then put the orange on a charcoal or cow dung fire and cook it: do not overcook the orange but cook it equally all over. Remove it from the fire and throw away the flowers and the dough. Wrap it in a clean cloth and eat it when required. If it is desired to keep it for many days, then remove the stuffing from the orange and dry it. Whenever it is required for eating, moisten it with rosewater and eat it. If an orange is not available, sew orange leaves together to form a basket, put the above-mentioned stuffing into it and cook it. A little is very good.

to create aromatic qualities none of them had to begin with, or to heighten inherent similarities. If you add jasmine and rose to a perfume, for example—ingredients that share some floral aromatics but are quite dissimilar—the characters of each flower merge, so that you can no longer discern either individually; yet together they contribute to a singular floral bouquet that is neither rose nor jasmine. Similarly, in food, we can ascribe to locking that wonderful sense of "beyond" that we experience when we put something truly delicious in our mouths. And it truly is "beyond," an effect of unity and transcendence that's greater than the sum of the parts. Or more precisely, the sum of the facets, which knit together to create a unique new flavor in which it is no longer possible to cleanly discern the individual elements. Nature, that master flavorist, creates locks within the context of a single ingredient, alchemizing hundreds of flavor molecules into the tastes we know as "strawberry" or "vanilla." A cook can create a further lock with as few as two ingredients, knowing that each contains multitudes—strawberry *plus* vanilla, for example. Or locking can be a highly orchestrated affair, with many ingredients contributing to a melded effect, as in a curry or a stew. At its most effective, locking creates a flavor that is only implied by the original ingredients—for instance, the phantom pine note created by the lock between grapefruit and rosemary. The ultimate lock is a satisfying finished, constructed flavor. It's like the fit of puzzle pieces, with all the facets that make them unique and distinct merging perfectly. Or maybe more apt, it's a bouquet that intoxicates with the mingled effects of all its elements, not simply a jumble of flowers.

Think about that cup of coffee we "cooked" in chapter 3. When you put coffee and milk together—even as a frothy cappuccino—you don't create a lock. The milk adds richness, dilution, and lightness to the dark, earthy flavor of the coffee, but the essential flavor of the coffee is unchanged. But if you add chocolate, the earthy and bitter facets of the coffee and chocolate will lock to create a new flavor, each ingredient surrendering its distinct traits to an indivisible whole. This particular flavor lock has become so popular that we gave it a name of its own: mocha.

Locking is on gorgeous and diverse display in cuisines around the world, from Indian curries to Mexican *moles*, the savory sauces characterized by multiple aromatic ingredients pounded together into a smooth, deeply layered mix—for example, Oaxaca's *mole negro*, which combines dried chilies, roasted tomatoes, pureed pumpkin seeds, oregano, and thyme with black pepper and clove in a foundation of bitter chocolate. In many cuisines this kind of sophisticated layering has a long history. Cooking from the Arab world is distinguished by centuries of combinations of sweet and savory ingredients for uniquely piquant locks, as in a dish known as *judhaba*, in which melon and other fruits, nuts, eggs, honey, and syrup were layered over a foundation of bread and placed in a clay oven under a roasting chicken or leg of lamb so that the fat and juices from the meat dripped down, infusing sweet with savory. "Loved by caliphs and ordinary folk alike, *judhaba* was sold by street vendors throughout the Muslim world, and few cuts of meat anywhere in Baghdad would have been roasted without a pan of *judhaba* under them," notes H. D. Miller in "The Pleasures of Consumption: The Birth of Modern Islamic Cuisine."

Roman recipes from Apicius's first-century cookbook *De re coquinaria* (On Cooking) demonstrate some elements of flavor locking, combining strongly flavored ingredients like vinegar, garum (fermented fish paste), honey, and raisin wine with potent herbs like oregano and mint. The use of spices in some of the recipes seems eerily modern, as in a dish of stewed pears baked in a custard with pepper, cumin, honey, broth, and wine. But for the most part, in the West, it wasn't until the eighteenth century that cooks began to manifest a sophisticated awareness of the effects they could create. As Jean-François Revel recounts in *Culture and Cuisine*, eighteenth-century French chefs "took balance to a new level, replacing the 'old-style cuisine of superimposition and mixture' (simple additions) with the 'new cuisine of permeation and essences' (subtle combinations)." The goal of the cook, as articulated in the introduction to a 1739 French cookbook noted by Michael Symons in *A History of Cooks and Cooking*, is to create flavors "so that no ingredient dominates the others and the taste of all of them comes through; and, finally, of giving them that unity that painters give to colours and of rendering them so homogeneous that all that remains of their diverse flavours is a fine and appetising taste and . . . an overall harmony of all the tastes thus brought together."

Creating flavor is not like constructing a building: it's not a matter of putting two static things together to create a third static thing. A specific mixture tastes a specific way at a specific moment, but flavor itself is dynamic, shifting with the addition of ingredients and the process of combining and cooking them. You should look at locking not just as a lucky accident that will (or won't) befall you when you make a dish but also as a concept that can help you predict, adjust, and learn from the interactions we call cooking, at every stage of the process. Locking, in other words, is a *tool*, the most important one in your kit. With deeper awareness of your ingredients and repeated experience, you'll learn to wield it with power and finesse, to anticipate and control the effects you're imagining. Your flavor memory—the sensory database of experiences that you're constantly compiling—will help you remember what works well and what doesn't, and will suggest new ideas and combinations. So locking is a conceptual tool that's useful when you're:

1. Imagining a dish or a flavor. Having locking in mind will help you predict what will happen as you combine and transform ingredients.
2. Cooking the dish. Locking will help you measure your progress in real time, allowing you to assess what's happened in a dish so far and how you need to adjust it to guide the flavor toward your goal.
3. Tasting the finished dish. Locking allows you to figure out what worked and what you can do better next time.

It's a cycle, in other words: A dish starts in your mind, but the first time you taste it—in fact, the first time you smell two ingredients and contemplate how they will work together—you have moved out into the field, so to speak, to engage with the actual dynamics before you. Tasting and smelling as you cook are ways to gather input to inform and edit where you are in the cooking process, and to assess the relationships between the different components. Learn to taste almost as if in slow motion—not the way you consume food at a meal. Smelling, tasting, and assessing the final dish feeds back into the first steps you'll take the next time you set out to cook. Locking provides a framework within which to situate data, so that your understanding grows and your cooking improves over time.

What does this process look like in practice?

Let's say you're in the mood for beef. As we've noted, the genesis of a dish isn't governed only by the imagination; practical considerations play a role as well. In this case, you're hosting friends, and you don't want to have to spend all your time in the kitchen once they arrive, and you don't want to spend a fortune, either. Luckily, both of these constraints point in the same general direction: a long, slow braise of an inexpensive cut of beef that can be made well in advance.

The cooking method, as we'll explore in depth in the next chapter, is a major player in facilitating a locking effect, whether it's the crushing and pounding that creates those unctuous *moles*, helping to unite ingredients that lie far apart on the flavor spectrum, marinating ingredients together for a period of time, or simmering them, as in a braise or stew. Or we may work against such wholesale melding, orchestrating ingredients so that, say, the flavor of a fresh herb added to a simmered mixture right before serving explodes in the mouth, like fireworks, or the flavors of a salad fire separately as you chew, creating exciting locks in the moment.

In this case you're opting for the melding of flavors encouraged by the long, slow heat of a braise. It's an ideal treatment for an inexpensive cut of beef like chuck. In fact, it's too bad that so much American beef cookery revolves around steaks, which are cut from the cow's less-used muscles. Steaks cook relatively quickly, and the pleasure in eating them is all about their tender, juicy texture and the flavor of the meat itself, putting a premium on the quality of the beef and how it has been aged, with just enough spice or sauce to heighten it but not distract. The tougher, cheaper cuts, from the animal's hardworking muscles, require long, slow, usually moist cooking at low temperatures to make them luscious and tender. In the process, they can acquire a depth of flavor that steak cannot match, with surprising and delicious layers of interlocking flavor introduced through rubs and spices on the outside of the meat, or infused into the cooking liquid.

Chuck is not only inexpensive but also readily available. You find it at the butcher counter already cut into big cubes, which is convenient. You put the meat in your cart and move your cart on. In the produce section, your eye falls on a pile of sweet oranges, showcased because it's the middle of winter and they're at peak season. Beef and orange is a flavor combination that's familiar to you from Chinese cuisine, and you muse on the syntax underlying that traditional pairing for a minute. You imagine how the sweet, floral facets of orange could wrap around and lock with beef's deep, gamey-sweet, umami-rich flavor, heightened by the braise. You decide that this will be the lock around which the dish you're making will revolve.

Beef and orange—sweet and rich—feels like it might need a bitter or sour element for balance. You're already familiar with the way that an acid like wine typically offsets the richness of the meat in a braise. But you also know from experience that wine becomes sweeter as it cooks, and as you extend the process in your mind, you imagine that the sweetness of cooked wine might lock with the sweetness of orange to make a broth that is flat and one-dimensional. (Locks don't always produce results that are more powerful than the sum of their parts—sometimes ingredients mesh in ways that diminish their individual force.) It occurs to you that you could employ a different kind of acid in this case, something capable of punctuating the beef-orange combination.

What about a high-acid, astringent black tea that can also act like a bitter, slightly peppery spice? You imagine what a lock between braised beef, sweet orange, and black tea would taste like. The sweetness of the orange and the richness of the beef could blunt the sharp, bitter edges of the tea, the way sweet, rich milk does in your mug. And the floral notes of the orange should brighten the tea, just as it brightens the beef. Your working understanding of locking is allowing your imagination to wander to such interesting new places!

Every decision matters. What kind of tea will you use? A straightforward black tea like English breakfast will work, but the smoked black tea called lapsang souchong, used judiciously, could contribute greater intensity and complexity. You smell the two side by side and decide that the lapsang souchong will make a more interesting partner with the beef and orange.

What else will go in the broth? Given all the intense flavors already in play, you think you will use water instead of beef broth as the braising medium, along with some orange juice, of course. Onions, carrots, and fennel will add a balancing sweetness, and in the right proportion—about a third the mass of the meat—won't overwhelm the character of the beef. The dish doesn't need much in the way of accompaniment. Bland white rice will absorb the broth and lend a subtle sweetness.

With all this in mind, you set to work. You brown the meat first, then remove it and add the vegetables, cooking them until they are also lightly browned and soft. All this browning adds depth to the dish, and softening the vegetables brings out a sweetness they will also contribute to the broth. You return the beef to the pot, add the liquids, and put the mixture to slow cook in the oven. You've decided to hold off on adding the tea until the end of the process.

When the meat is tender, you taste the broth. It is rich but not overly dense, and lightly sweet, but the orange has all but disappeared under the beefy flavor. So you decide to add not only the tea but some orange peel as well, to get some of its more aromatic qualities into the liquid. You strain the cooking liquid into a bowl and steep the tea and orange peel in it for just a few minutes, until the liquid is flavorful but not too bitter. Then you strain the broth back into the pot, discarding the orange peel and tea leaves. You cover the pot again and let it sit for about fifteen minutes, so that the flavors have a chance to meld. You taste again. You think that the orange and the beef and the tea could use one more element to bind them, a concentrated flavor that can stand up to the other elements, with a sweetness that can support the orange. What about a dried fruit, with its concentrated sweetness? You consider dates, but you think they might be overwhelmed by the smoky tea. What about prunes, which have a bit of acidity and a deep, umami quality of their own? You taste a prune and consider—the facets seem like they will match the deep, dark flavors of the tea and complement the orange as well. You add them to the pot and let the dish sit until your guests arrive.

Just before serving, you taste the dish again. It has melded so smoothly that it needs a little spark, but just a little. Then, after you spoon the stew over rice in individual bowls, you decide to zest some orange over each, and also give each a few turns of black pepper to punctuate the smoothly blended flavors of the braise, treating the pepper like a true spice. Then you sit down to your own portion. You savor the scent of the steam rising from it, then taste it. What do you think? It's pretty terrific. The lock of beef, orange, and tea is interesting and original, mediated by the prunes in just the way you'd hoped, and diluted by the rice. What could be better? The smoky flavor of the lapsang souchong is a little too strong, and it buries the orange. Next time, you think, you will use less tea and more orange peel, and maybe a little more freshly squeezed orange juice at the end, or even a shot of rice wine vinegar to punctuate the richness of the dish.

As you eat, your mind wanders on. You could approach the basic idea in a different way. What might happen if you *did* use dates? That feels like it might create a more neutral palette, so maybe in that case you'd use regular black tea instead of smoked. You still want to create some layers of flavor, though, so you consider adding chili for spice, and cinnamon and star anise for complexity. And so you are back to step one.

Braised Beef with Black Tea and Orange

SERVES 4-6

3 POUNDS CHUCK, IN LARGE CUBES

SALT AND FRESHLY GROUND BLACK PEPPER

4 TABLESPOONS VEGETABLE OIL

2 MEDIUM ONIONS, CHOPPED

2 LARGE CARROTS, PEELED AND CHOPPED

1 FENNEL BULB, CHOPPED

1 CUP FRESHLY SQUEEZED ORANGE JUICE

3 CUPS WATER

PEEL OF 2 SWEET ORANGES, IN LONG STRIPS

2 TABLESPOONS BLACK TEA LEAVES

2 CUPS PITTED PRUNES

COOKED RICE

FRESHLY GRATED ORANGE ZEST

Preheat the oven to 300 degrees.

Season the beef generously with salt and black pepper. Heat the vegetable oil in a wide, heavy, ovenproof pot and brown the meat on all sides. Take care with this step, because you will be cooking with water, and the browning of the meat is a crucial element in flavoring the broth. Add the onions, carrots, and fennel to the pot and continue to cook over medium heat until the vegetables are lightly browned and softened, about 5 minutes. Add the orange juice and water, and bring to a boil. Cover and cook in the oven until the beef is tender, about 2 hours.

Strain the broth into a nonreactive saucepan, returning any vegetables to the pot with the beef. If there is excessive fat in the broth, skim it off and discard. Add the orange peel and tea leaves to the broth and bring it to a boil. Remove from the heat and let it steep, tasting often, until it is flavorful but not too bitter, about 5 to 8 minutes. Strain the broth back into the pot, discarding the orange peel and tea leaves, then add the prunes. Cover the pot and let it sit for about 15 minutes, to meld the flavors. You can do this much in advance, and in fact, if you make it the day before, it will only get better as it sits. If preparing the stew more than a few hours in advance, refrigerate it.

When ready to serve, bring the stew to a simmer. Ladle the stew over rice, zesting some orange and grinding a little black pepper over each serving.

☞ Tea—pouring hot liquid over an aromatic material, letting it steep, and using the resulting liquid—is a great way to extract flavor without substance. Such infusions can be used in soups, broths, as cooking liquids, or as drinks. An infusion can be introduced at the beginning of a cooking process, in which case its flavor will meld with the other ingredients, or, as here, at the end, where its flavor will remain more distinct. Try other kinds of combinations, too, like chicken soup scented with green tea, or rice pudding infused with jasmine.

{ BURYING }

What creates the magic sense of transcendence that we call locking isn't just the right combination of ingredients, it's the right *proportions* of the right combination of ingredients. The beauty of a bouquet of flowers doesn't depend only on a felicitous combination of colors and shapes and scents; it depends on the decision to highlight some elements and relegate others to supporting roles, taking into account the nature of each. If one of the mix is particularly brightly colored or oddly shaped, you will probably use less of it, incorporating it in a way that it energizes the whole without dominating.

As we explored in the previous chapter, some of the most exciting ingredients we get to work with to create flavor are the most intense—the herbs, spices, flowers, and citruses. When we work with them, even in complex, multi-element blends, we generally decide to highlight one or two. The relative dynamics of flavor—the ability of some elements to control (or, if not controlled, to overpower) others is what Mandy calls burying. Distilled to their simplest relationships, ingredients—like essences in a perfume—are either locking with or burying one another. As Mandy teaches her perfume students, the two phenomena are the most important conceptual tools in composition. Pack them in the top of your culinary kit and put them to work at all times!

Like locking, burying can occur unintentionally and infelicitously, eclipsing an ingredient that can't stand its ground under an onslaught of more assertive elements. But also like locking, burying can be invoked deliberately and artfully. Used in a thoughtful way, it can push a strong flavor just far enough down that it has a subtle, transformative effect on the whole, to achieve a balance of different elements. For example, in the beef and orange recipe, the lapsang souchong is a strong flavor that needs to be controlled—to be buried. You already use burying without thinking about it when you add a grind or two of black pepper to give a lift to your salad or zip

This recipe for pickling mint leaves from al-Baghdadi's thirteenth-century cookbook is a textbook case of burying—in this case, the spicy herb:

Take large-leaved fresh mint and clean the leaves from the stalks, then wash them and dry them in the shade. Sprinkle aromatic herbs on them. If you like, add celery leaves and peeled cloves of garlic. Put it in a clean glass jug, cover it with good vinegar and colour it with a little saffron. Leave it until the leaves absorb the sourness of the vinegar and its sharpness is cut; and use it.

to your soup without dominating them. Intentional burying is never absolute. You want to bury the strong flavor of sage in a lamb stew just far enough that it remains mildly present as a seasoning, not overbearing. When strong ingredients are buried, sometimes different facets of them are revealed—as with the resinous and floral notes of sage, which, when its intensity is properly buried, rise to the surface.

Burying can also be a way to remedy flavors that have developed in an unpleasant way. Maybe you've accidentally scorched the bottom of your stew a bit; adding a little more water or broth or some neutral grains can dilute the flavor and help bury the acrid taste, as can a squeeze of lemon juice. Introducing a bit more fat in the form of a pat of butter might help round out the bitter edges. Burying is a way not only to accent the flavors you want but also to push down flavors you didn't want or expect.

Proportion

Both locking and burying depend on the mastery of proportion. Proportion is how much you use of your ingredients in relation to one another. Proportion is an essential consideration in the composition of a dish, down to the amount of chicken on the plate relative to the amount of mashed potatoes, or the amount of salt relative to the amount of water in the broth. Or the amount of dressing compared with the amount of greens. Even the proportion of the dressing itself matters. A vinaigrette of one part vinegar to three parts oil might be perfect as the basis for a subtle marinade or accompaniment for grilled meat, but too flat for dressing a salad. Proportion can make a dish sublime, or ruin it.

As with that flower bouquet, it's rare that exactly equal proportions create the transcendent flavor we're looking to achieve. Sometimes it does happen—usually when we're working with ingredients of similar character, texture, and intensity. In chapter 3, we made a puree that resulted in a lock between two very similar vegetables, carrots and sweet potatoes. Root vegetables work particularly well this way, as do members of the nightshade family (tomatoes, eggplants, and such) and the Cucurbitaceae family, which includes not only squashes and gourds but cucumbers and melons as well.

Cucumber-Melon Soda

SERVES 6

Aside from the fact that they tend to ripen around the same time in the summer, we tend to think of cucumbers and melons as far apart—savory and sweet. But they share both a family name and a sweet, watery character, with light green, floral top notes and a similarly middling intensity. Here these ingredients are combined in equal proportion, and the facets lock into a green, sweet, floral flavor that is neither one nor the other. They are both flat, so a touch of lime lifts the mixture. Spearmint provides a spark, playing against the herbaceous, floral notes of the principal ingredients, but with a rounder, more embracing feel than peppermint. With the bright and sparkling quality of the combination, the effervescence of a soda is a natural vehicle. The soda also relieves the thickness of the melon juice—texture matters, even in drinks. Do make sure you are using sweet, aromatic cucumbers and ripe melons—muskmelons like Ha'Ogen and Charentais, with their faintly funky, animalistic "musk" melon note, are particularly great if you can find them.

½ POUND RIPE MELON (PREFERABLY MUSKMELON), PEELED, SEEDED, AND CUT INTO CHUNKS

½ POUND RIPE CUCUMBERS, PEELED, SEEDED, AND CUT INTO CHUNKS

2 TABLESPOONS CHOPPED MINT

4 TABLESPOONS FRESHLY SQUEEZED LIME JUICE

1 TABLESPOON PLUS 2 TEASPOONS SUGAR

CLUB SODA

Combine equal amounts of melon and cucumber in a blender with a good handful of mint. Blend the mixture until smooth, then strain it into a bowl set into an ice bath to chill the liquid quickly. Add the lime juice and sugar to taste, then add a bit more, as the club soda will dilute the flavor.

As with the Strawberry and Rose Drink (page 140), adjust the ingredients to create the sweet-sour balance and concentration that feels right to you. Fill the glasses halfway, over ice if desired, and top with an equal amount of club soda.

☞ With the addition of gin or vodka, this makes a great midsummer variation on a Mule.

THIS DELIGHTFUL SODA NOTWITHSTANDING, felicitous combinations of ingredients in equal proportion are the exception, not the rule. Most often, the combinations that please us tend to tilt in one direction or another. As with painting or music, we usually want maybe one or two elements to be highlighted, the others to do the highlighting, or to round them out or layer under them. Learning to create flavor requires developing a feel for which proportions of combined ingredients will be pleasing. The perfumer and perfume educator Jean Carles developed an exercise that Mandy has adapted for teaching her perfume students about proportion and intensity. To each of three small vials containing the same small quantity of perfume alcohol she has her students add a total of ten drops of the same two essences. In the first vial the proportions are seven drops of one essence to three drops of the other; in the second, they are equal, five drops of each; and in the third, they are tilted the other way, three to seven. An odor intensity study of black pepper and vanilla, for example, would be three drops vanilla and seven drops black pepper in the first vial, then five drops each, then seven drops vanilla and three drops black pepper. As the proportions shift, the perfumer discovers not only where the two aromas balance each other perfectly—their relative odor intensity—but also the aesthetic sweet spot where the perfumer likes the blend best. The two are not necessarily the same. In fact, they are rarely the same—almost always, one of the tilted proportions is chosen: seven and three, or three and seven. Again, we generally prefer an appealing lopsidedness to a rigid evenness.

Orange Lemonade

This recipe is the food equivalent of a Jean Carles study, teaching you how different proportions of the same two ingredients can have radically different effects. As we know, two ingredients from the same family—citrus, for instance—do not necessarily have equal intensity. Despite its sourness, lemon juice (as opposed to peel) is low on the intensity scale and thin in character, which is why it is often used to adjust other ingredients (as we'll explore in chapter 8). Because of this quality, it can add significant sweet-sour effects while its flavor remains in the background. Here, the addition of a little lemon juice to freshly squeezed orange juice helps moderate the sweetness of the orange juice and gives it a full, round, mouth-filling flavor without stealing the spotlight. The addition of water stretches out the flavor and makes the drink more like lemonade and less like a glass of orange juice.

FRESHLY SQUEEZED ORANGE JUICE

FRESHLY SQUEEZED LEMON JUICE

SUGAR (OPTIONAL)

Dilute the orange juice with half as much water. Begin adding lemon juice a tablespoon or two at a time, tasting as you go. Stop when the mixture reaches an appealing balance. If the orange juice isn't sweet enough, add a little sugar. (If you want to turn this into a true Carles study, keep adding lemon juice until you get to the point where the lemon flavor dominates and the orange recedes, tasting as you go to appreciate where the tipping point lies.)

☞ Citrus juice contains only a little of the essential oil found in the peel, so if you want to amplify the aromatics, zest some citrus rind into the finished drink, refrigerate for at least an hour, and then strain. The aromatic intensity will have increased considerably.

Herb Salad

SERVES 6

This recipe is an exercise in the discipline of proportion. Although you are starting with a bunch of each herb, you must resist the temptation to use all of any of them just because you have them. You must follow your taste to create a balanced mixture, taking account of relative intensity and using only so much of a given ingredient as will create the effect you want. The basil, parsley, and chervil are lowest on the intensity scale: use three parts of each. Tarragon is sweet but has a driving anise flavor that can overwhelm the elements around it, so it gets only two parts. And chive, with its oniony flavor, is so sharp and different that it will stand out in the mix and needs to be muted; it gets only one part.

The frisée plays an important role in flavor here, too—not only by contributing its own mild bitterness but also by keeping the herbs from clumping together and creating too high a density of flavor.

½ BUNCH FRESH BASIL

½ BUNCH FRESH TARRAGON

½ BUNCH FLAT-LEAF PARSLEY

1 BUNCH CHIVES, CUT INTO ¼-INCH PIECES

½ BUNCH FRESH CHERVIL

1 HEAD FRISÉE

2 TABLESPOONS OLIVE OIL

2 TABLESPOONS CHAMPAGNE VINEGAR

SALT AND FRESHLY GROUND BLACK PEPPER

Wash and dry the herbs and pick the leaves off their stems. Combine all the herbs and the frisée in a mixing bowl, adjusting the quantities until you have a balance you like. Dress the herbs lightly with the oil and vinegar, and season with salt and black pepper.

The Herb Salad is too intense to stand on its own, but it makes a great accompaniment to grilled meats and fish, roasted beets, or new potatoes.

Have you exercised proper restraint and used only as much of the herbs as the salad needed? Good! Your reward is this great use of the leftovers:

Remove all the really thick stems, then roughly chop the herbs. Place them in a food processor or blender and add a squeeze of lemon, some salt and freshly ground black pepper, and just enough neutral-tasting oil to blend them all together until smooth and bright green. Taste and adjust the seasonings. Transfer to a jar and store it in the refrigerator, then use it anytime you want a quick way to enliven bland base ingredients—rice, potatoes, pasta. A dollop is great in soups and stews as well.

Intensity

As we've said before, nature is the ultimate flavorist. Nature's palate has perfect symmetry, but the composition of its ingredients is anything but symmetrical. As we saw in chapter 2, ingredients vary enormously in intensity—most meats and grains and vegetables are of relatively low intensity, but members of the onion family are of high intensity. Yet even among similar kinds of ingredients, intensity—and therefore flavoring power—is not uniform. Lamb is more intense in flavor than pork, and a bitter vegetable like radicchio dwarfs a mild lettuce.

As we saw in the previous chapter, even the ingredients we most often turn to in order to create flavor—herbs and especially spices, both rich in essential oils—are among the most intense. Medieval cooks, who, to judge from their recipes, used an astonishing array and amount of spices—ginger, pepper, cinnamon, saffron, cloves, mace—nevertheless tempered their potency with the addition of starch and creamy sauces.

Yet as we know from the flavor compass, even the most highly flavorful ingredients vary enormously in their intensity.

That knowledge is essential to building flavor. The ingredients at the top of the intensity scale are spicy chilies, fermentations like fish sauce and kimchee, herbs like rosemary and lavender, spices like cinnamon and cumin. In the perfume studio, Mandy calls these high-intensity ingredients "accessory notes." In the kitchen as in the perfume studio, they must be used in a fundamentally different way. They can have an alchemical effect on the other ingredients around them, transforming the whole without imposing their own identity. But they also have the power to take over and ruin a blend in a way that less intense ingredients do not. They must be buried so that they alchemize the flavor rather than obliterate it.

The black tea in the Braised Beef with Black Tea and Orange (page 152) is an accessory note. So is the black pepper in the following recipe. The introduction of black pepper into ordinary vanilla ice cream may seem esoteric, but it makes sense to the imagination. For all their differences, the two ingredients share floral facets that wrap around each other when combined, the dusty, earthy, spicy pepper lifting and showcasing the deeply perfumed, sweet heaviness of the vanilla rather than claiming center stage.

Vanilla—Black Pepper Ice Cream

SERVES 4–6

In this recipe, cream and sugar provide a medium that softens and connects the disparate flavors of vanilla and black pepper. Note the absence of eggs, which feature in many ice cream recipes. Eggs have a distinctive flavor, and omitting them allows the flavor of the other ingredients to be more clearly expressed. Black pepper essential oil works beautifully in this recipe, as it has all the flavor of black pepper without the spiciness. You can also use whole black peppercorns by steeping them in the warm cream, but they will express less of the floral notes and more of the spiciness, which you may find disconcerting in a dessert. Take care not to eclipse the vanilla, though. It should be the dominant note, with the black pepper buried, relegated to the role of seasoning for the vanilla.

2 CUPS HEAVY CREAM

2 CUPS WHOLE MILK

¾ CUP SUGAR

1 VANILLA BEAN, SPLIT

3–4 DROPS BLACK PEPPER ESSENTIAL OIL (OR 2 TEASPOONS WHOLE BLACK PEPPERCORNS)

In a medium saucepan, heat the cream, milk, sugar, and vanilla bean to just below a boil and turn off the heat. Let the mixture stand for 30 minutes, then remove the vanilla bean and scrape out the seeds into the liquid, discarding the pod; the seeds contain most of the flavor in a fresh vanilla bean.

If using the black pepper essential oil, add it one drop at a time, tasting after each addition, as it is very strong. Stop when the vanilla is still on top but the black pepper adds a dusty, earthy note to the background (think of a 7:3 Carles ratio). If using peppercorns instead of essential oil, add them once the vanilla bean is removed; reheat the liquid until it is very warm and let stand. Taste the infusion as it sits, then strain to remove the peppercorns when the desired intensity is reached.

When the mixture is cool, transfer it to the refrigerator and let it chill for a couple of hours before freezing it in an ice cream maker.

☞ Vanilla and black pepper can make a great savory combination as well. Blend vanilla bean and freshly ground black pepper into a neutral-tasting vegetable oil, then let it infuse for a day. Strain and use on seared scallops or duck breast.

As we've mentioned, burying sometimes occurs unintentionally, eclipsing an ingredient that can't stand its ground under an onslaught of more assertive elements. The greater the difference in intensity between ingredients, the greater care you must take to give the more delicate ingredients the space to shine. Use burying artfully, to bring an intense flavor down to an accent, not an obstacle.

Roasted Delicata Squash with Pumpkin Seed Oil and Togarashi

Delicata squash is well named—it has a thin skin, edible when roasted, and a delicate, floral flavor, without the depth of most other squashes. These unique qualities require a thoughtful approach. A puree would be too light in flavor, but the firm texture of the squash makes it good for serving as a side dish. Roasting the squash brings out an earthy nuttiness that could be amplified with a bit of store-bought pumpkin oil (assuming you're not so dedicated a cook that you decide to press oil out of the squash seeds!). Now the squash has more presence, but it still needs a top note compatible with its earthy sweetness. Citrus would be distracting, and would bury the subtle nutty notes you have worked to emphasize. How about finishing the flavor by tossing the roasted squash in something salty and spicy to give it an exterior punch? For the spicy element, the Japanese chili powder shichimi togarashi *is a mixture of several spices, including chilies, that make it multidimensional rather than simply just spicy. But controlling proportion is key—don't overdo the seasoning, or it will bury the squash flavor.*

DELICATA SQUASH

VEGETABLE OIL

PUMPKIN SEED OIL

SHICHIMI TOGARASHI

SALT

Preheat the oven to 400 degrees.

Halve the squash, scoop out the seeds, and cut it in thick crosswise slices. Toss the slices with enough vegetable oil to coat them, then lay them out on a baking sheet lined with parchment paper. Roast for about 20 to 25 minutes, until the squash is tender and lightly browned. Transfer the squash to a serving bowl and toss it with a couple of tablespoons of pumpkin seed oil, *shichimi togarashi*, and salt.

☞ This recipe creates a nutty, deep, sweet, spicy flavor. If you want something brighter, grate some lime or lemon zest over the dish to finish it. Or for a more elegant version, try buttery hazelnut oil, sweet orange zest, and freshly ground black pepper.

Recognizing trace elements of one ingredient in another can suggest that they might go well together. But just because they are traces doesn't mean they aren't strong—in fact, strength is what gives them the power to be transformative. So when you make those connections between facets, tailor the proportions accordingly. For example, if you taste basil with an open mind, you will find cinnamon notes. This is because both cinnamon and basil contain eugenol, which allows basil to lock with cinnamon. When they are combined, however, as in the following dish, the cinnamon needs to be buried, or it will block out the basil, which is fresh and bright and meant to sit on top, a cheerful aroma to greet the diner before the first bite.

Duck Breasts with Endive, Honey, Cinnamon, and Basil

SERVES 4

Duck breast is a slightly gamey, extremely flavorful, and versatile ingredient that has more in common with red meat than with chicken and most other poultry, and requires stronger accompaniments. It wants to be cooked to medium-rare, rather than all the way to doneness, which will render it dry and tough. Duck breast's rich, fatty qualities benefit from sweet, sour, bitter, and spicy counterpoints: it needs something dynamic as balance. Strong herbal accompaniments like rosemary, thyme, and parsley are fine—like steak, duck takes well to simple treatments. Almost all spices work well with duck, too, but the sweet spices have a special affinity. Cinnamon would be a good partner, but it needs a top note. Basil? At first thought it seems like an odd combination, because cinnamon belongs to the world of desserts, to "sweet spices," and basil is firmly in the savory camp. But as noted, their shared facets suggest this pairing.

The duck itself is cooked gently in a pan, until the skin is browned and the flesh medium-rare, and then glazed with a mixture of honey, vinegar, cinnamon, salt, and black pepper. Honey and vinegar are a great combination. The honey rounds out the vinegar, making it less aggressive, and the vinegar takes away some of the sweetness of the honey, leaving the floral, aromatic notes on the top. Endive lends a welcome bitter note.

As advised, proportion is crucial to keep the cinnamon properly buried. Use freshly ground cinnamon in the honey-vinegar mixture if you can, but just a pinch, a suggestion, a whisper. Then sprinkle some basil over the top, which will envelop and enliven the deep flavors underneath.

4 DUCK BREASTS

SALT AND FRESHLY GROUND BLACK PEPPER

4 TABLESPOONS HONEY

3 TABLESPOONS CHAMPAGNE VINEGAR

½ TEASPOON CINNAMON, FRESHLY GROUND IF POSSIBLE

1 TABLESPOON VEGETABLE OIL

1 HEAD ENDIVE, SLICED

HANDFUL FRESH BASIL LEAVES, CHOPPED (OR USE A SPRAY OF ESSENTIAL OIL)

Season the duck breasts with salt and pepper and let stand for 20 minutes. Combine the honey and 2 tablespoons of the vinegar with the cinnamon and lots of salt and black pepper. The flavor should be very complex, the earthy, dusty black pepper marrying well with the cinnamon but staying just underneath.

Place a sauté pan over medium heat. Add the duck breasts and cook, turning often, until the skin is nicely browned and the meat is done to your liking, medium rare-to medium, still pink inside and tender to the touch. (You may also broil the duck.) Remove the duck breasts from the heat, brush with the honey mixture, and set aside to rest.

Return the sauté pan to the heat with the vegetable oil. Add the endive and some salt, toss the endive a few times to wilt it, and then add the remaining vinegar, tossing a few more times to incorporate.

Divide the endive among plates. Slice the duck and lay on top of the endive, brush with the honey mixture again, and sprinkle with the basil. Drizzle any remaining honey sauce around.

👉 The kind of sweet-sour sauce in this recipe works especially well with game birds. The sweetness locks with the gamey richness, and the vinegar balances the dish. For a less floral, deeper flavor, replace the honey with caramel (half sugar and half water cooked until deeply browned).

As in locking, and as we explore further in the next chapter, the cooking method plays an essential role in burying. How the accessory notes are treated—fried, infused in a liquid, or minced—is key to creating a harmonious combination. These three recipes showcase different ways of controlling the intensity of a powerful herb, rosemary.

Potatoes Fried with Rosemary

Potatoes, with their mild, sweet, earthy, deep flavors, make a good base for other layers of flavor. This dish is based on a traditional Spanish recipe, patatas bravas, *in which the potatoes are boiled, then smashed and fried, giving them a wonderful textural dimension. They look cool too!*

But frying the potatoes in their skins influences more than their texture. It gives them a deep, complex flavor. Even so, chopped raw rosemary sprinkled on top would still be overpoweringly green and resinous. But frying the rosemary along with the potatoes tames its character, softening its aggressiveness and turning the leaves crisp and delicious. In turn, the rosemary flavors the oil while it cooks, scenting and boosting the potatoes.

SALT

12 NEW POTATOES

OLIVE OIL OR VEGETABLE OIL

2 TABLESPOONS ROSEMARY, LEAVES ONLY

Simmer the potatoes in salted water until tender. Drain and cool.

Smash each potato with the back of your hand until it flattens and the skin bursts. Be aggressive—there's no wrong way to do it.

Heat the oil to a depth of half an inch in a heavy, wide pan, such as a cast-iron skillet. When a small piece of potato sizzles on contact, the oil is ready. Add only as many potatoes as will fit in the pan in a single layer, cooking in batches if necessary. Cook over medium-high heat until browned, then flip them. Sprinkle generously with the rosemary and fry for 1 or 2 minutes more, crisping but not burning the rosemary leaves.

Remove the potatoes and rosemary to a bowl, toss them with a generous amount of salt, and serve the potatoes in a jumbled pile.

☞ This method also works well with sage, as in the Brussels Sprouts Fried with Garlic and Sage (page 131).

Potato Puree Infused with Rosemary

Another way to bury the rosemary is with a controlled infusion, from which you can remove the herb the moment the balance of flavor in the liquid is perfect. Used this way, rosemary shows a sweet, floral side, and the pine elements come out strongly. The Yukon Golds have a sweeter, warmer flavor than russets, and those facets lock with the floral notes of the rosemary. The milk stretches the intense flavors and leaves a lightness on the palate. The potatoes don't need much else, only a bit of butter to create a bridge between the saplike rosemary flavor and the earthy potatoes.

SALT

2 POUNDS YUKON GOLD POTATOES,
   PEELED AND QUARTERED

4 OUNCES BUTTER

1 CUP WHOLE MILK

2 LARGE ROSEMARY BRANCHES

Bring a large pot of well-salted water to a boil, add the potatoes, and simmer until they are tender. Drain them in a colander and discard the cooking liquid.

While the potatoes are draining, put the butter and milk with the rosemary and a generous pinch of salt in a small saucepan and place over medium heat. When the mixture is just below a simmer, remove from the heat. Taste. If the rosemary flavor is strong, remove it. If not, let the mixture stand a minute or two more. The flavor has to be just a little high, because the potato will balance it.

In a mixing bowl, crush the potatoes with a fork. Stir in the hot liquid. Season with salt.

 Thyme works well here also, making a sweeter, more gentle infusion.

Romaine Salad with Parmesan and Rosemary Vinaigrette

American-grown rosemary can be an especially difficult ingredient to work with raw, as it tends to be a particularly harsh variety of the herb. In Sicily, the rosemary tends to be sweet, tender, and mildly flavored, and cooks there scatter whole sprigs of it over carta di musica, *a flatbread. When Italian restaurants in the United States try to re-create the dish verbatim, however, the force of the herb unpleasantly buries the rest of the ingredients and renders the dish unpalatable. So judge proportions first and foremost by the quality of the ingredients at hand.*

In this case, rosemary is chopped into tiny pieces and mixed with vinegar and oil as the base for a simple but interesting salad. Sometimes accessory notes, used judiciously, can move a familiar combination in an exciting way. In this simple salad of romaine leaves, grated Parmesan, and croutons, a very small amount of finely minced rosemary in the vinaigrette changes the familiar flavors to something subtly new. On its own, the raw rosemary would be overpowering, but here it locks with the Parmesan and vinegar. The balance to the salad is very important—lots of cheese, which brings up umami and adds fat, just enough vinegar to brighten, and a tiny bit of rosemary. Neutral croutons tone down the intensity and add crunch.

2 TABLESPOONS RED WINE VINEGAR

1 TABLESPOON FRESHLY SQUEEZED LEMON JUICE

4 TABLESPOONS FRUITY OLIVE OIL

PINCH MINCED FRESH ROSEMARY

SALT

4 HEADS ROMAINE, LEAVES SEPARATED

½ CUP FINELY GRATED PARMESAN

½ CUP CROUTONS BAKED WITH OLIVE OIL AND SALT

FRESHLY GROUND BLACK PEPPER

Mix the vinegar, lemon juice, and olive oil with the rosemary and salt; add the rosemary a little at a time—you can always add more, but you can't take it out once it's in there. The vinaigrette should be sharp with vinegar but slightly mellowed by the lemon, and the rosemary should be a refreshing background note, barely audible.

Toss the romaine, Parmesan, and croutons with some of the vinaigrette and salt and black pepper. Add enough vinaigrette to coat the leaves well and make the salad bright and vibrant.

☞ Rosemary, minced very fine and used in small amounts, can have a magical effect on all kinds of preparations: cooked with onions and garlic in a white bean soup; marinated with olive oil and flank steak before grilling; added to a lemon butter sauce for a strongly flavored fish like salmon.

As you can see, cooking methods have a vast impact on flavor that we have only touched on here. Choices made in how the ingredients are prepared can accentuate some facets, diminish others, and create entirely new locks and other dynamics, a topic we'll explore in greater depth in the next chapter.

Seven

THE RAW AND THE COOKED . . . AND THE BAKED, STEAMED, AND GRILLED

For once, nature does things less well than we do. Our savoir-faire magnifies the given, which belongs to a suborder when raw. The aroma of roasted coffee early in the morning makes our muscles and skin quiver with delight; the smell of roasting meat, which verges on that of burning meat, delights our spirits—although rather less so than caramel, mere sugar until it meets fire. . . . Fire fuses many things together. The raw gives us tender simplicities, elementary freshness, the cook invents coalescences.

MICHEL SERRES, *THE FIVE SENSES*

ometimes the chef regards the perfumer with envy. Yes, perfume ingredients must be gathered and blended with infinite care, as they represent the purest expression of the relationship between aromatic ingredients. No nuance will be missed by the wearer. On the other hand, all the ingredients arrive at the perfumer's studio in more or less finished form, ready to combine with other ingredients. They will affect one another as they are blended, but they are little changed by their medium.

In cooking, however, flavor is embodied in solid matter, in ingredients that must be processed—peeled, pounded, cut, chopped, ground, and so on—and most of which will furthermore need to be cooked before, or as, they are combined.

Even the simplest step in altering an ingredient is part of the cooking process. The second that you cut an ingredient, you have started to transform it. Cooking, remember, is the transformation of edible things from one state to another. Good cooking is about controlling this transformation. Sometimes that means exercising restraint—it could be argued that a perfect peach can only be diminished by any transformation at all, although even allowing that peach to become perfectly ripe in the first place is itself part of the process. Or it might mean choosing from an array of applications of processing and applying heat, from the softening addition of salt and vinegar to the subtle melding of flavors brought about by long, slow cooking. Cooking is not about simply mastering a technique, but also about considering the right cooking method as an integral component of flavor. In a sense, cooking is an ingredient in its own right.

It is useful to think of cooking methods on a spectrum or progression, from raw to soft-cooked to hard-cooked. On the less-touched end of the spectrum would be that perfectly ripe peach, which wants to be eaten out of hand, with no seasoning or anything else that might interfere with what is already perfect flavor and texture. On the other end would be a piece of meat, or vegetables of lesser quality, which might require a lot of seasoning and manipulation to achieve deliciousness.

Most of cooking, however, sits between those two extremes.

❨ UNCOOKED ❩

Raw

Understanding an ingredient starts with encountering it raw. Ideally—except where this might be dangerous, as with, say, chicken or pork—tasting it raw. Yes, the smooth skin of a turnip or the bright eye of a fish is an indicator, as is the sweet smell of a peach (or lack thereof). But even a turnip that smells delicious can be metallic and fibrous to the tongue, crying out for cooking. Or maybe it's delightfully crunchy and sweet enough to eat raw. Taste it!

Yet even raw ingredients are altered—cooked, you might say—by process. For example, the green outer skin of a young artichoke is bitter and unpalatable. But once the skin is removed, the pale yellow flesh underneath is sweet, grassy, and nutty, and makes for a delicious salad, shaved thin. Peeling is preparation—it is cooking.

So is cutting. Cutting—the act of consciously changing the shape of an ingredient—allows the cook to control proportion, cooking time, presentation, texture, and the way that ingredients integrate. How ingredients are cut makes a big difference in how they taste. Lemons cut into large chunks are intense, dominant. Slivered into tiny pieces, however, they can provide relief from richness, or add a subtle piquancy to a sauce. Likewise, whole leaves of mint create big pops of flavor, whereas thinly sliced pieces stirred into a dish meld with the ingredients around them, creating more of a backdrop. When we chop an herb, in fact what we are really doing is proportioning the essential oil by controlling the shape of the ingredient in which it is contained. Different ingredients require different treatments, depending on their texture, and different treatments of the same ingredient result in subtle—or sometimes dramatic—differences in flavor. Sometimes, as with spices, or with sauces like pesto, the ingredient needs to be smashed or ground to reveal a certain flavor. An ocean divides a classic aioli—the garlic mayonnaise built on the powerful, pervasive flavor of garlic pounded to a puree with herbs—from a mayonnaise that has been mildly flavored with a bit of minced garlic.

Using a sharp knife to cut, slice, or chop is essential when you want to preserve the integrity of an ingredient as much as possible—when you want, for example, to have small cubes of ingredients whose crisp, clean, and distinct flavors release with intense little bursts in the mouth rather than being smoothly melded before they reach it, as in, say, an Israeli salad of diced onions, cucumbers, and tomatoes. A sharp blade will cleanly cut through cell walls, while a dull one will smash a wider swath of the ingredient, releasing its flavor prematurely. There is a use for crushing through cell walls—it can create a more intense and more merged flavor, as in a raita in which cucumber is grated into yogurt, creating a medium in which spices, herbs, and yogurt can readily mingle. But you want that choice to be deliberate.

The right tools not only make cooking easier but make a real difference in flavor as well. Good-quality knives, with blades that can be sharpened to a fine edge, are the most versatile. A mandoline is fast and easy to use, and it can slice vegetables and fruits more thinly and consistently than a knife can. A well-sharpened cleaver is a versatile tool for both cutting and smashing. A mortar and pestle makes it possible to grind spices smoothly without heating them, as an electric spice grinder does, thereby preserving more of their flavor and their essential oils. When grinding or crushing ingredients, it's also easier to control texture in a mortar and pestle, stopping just when the desired balance of smoothness and coarseness is reached rather than leapfrogging to the smooth puree a food processor creates. That's why a handmade pesto or aioli is superior not simply in texture but also in taste to a machine-made one.

Cutting by hand allows you greater control than machines do. As noted, it keeps ingredients cool—so closer to their freshest state. But more than that, it allows you to stay in touch with what is happening at each step of the process, how the flavor and aroma are changing. The feedback that comes from direct experience with materials helps build your flavor intelligence.

In some dishes, "raw" is the star, and in these dishes, the spotlight on the method of processing the ingredients is especially intense. Raw fish and meat are sliced thinly and dressed lightly as carpaccio, or finely diced or minced—not ground!—as tartare. Or think of the exquisite attention to slicing that goes into a perfectly presented assortment of sushi. Raw vegetables like onion and cucumber add snap to a sandwich, salad, or main dish—and with loving treatment can take center stage. Shaved thinly and seasoned with a vinaigrette, raw vegetables make a great salad.

Shaved Raw Squash Salad with Mint

It might seem strange to create an entire dish around squash, when it's so often used as a side dish or filler, something cheap to sauté with more fabulous summer produce like tomatoes or peppers, or to throw into a pasta or stew. But squash, which gets its name from the Native American word for "to eat raw," makes a wonderful salad, especially when it is very young. Younger generally means sweeter and more tender. Look for small squash with shiny, unblemished skin and a firm texture.

Squash is not a strong flavor, so accompanying ingredients must be chosen with care not to overwhelm. It's also a flat flavor, so it needs a bit of acidity and an herb for brightness. If it's midsummer, basil comes to mind. For a more traditional flavor profile, lemon juice gives the basil a soft, round flavor, perfect for a salad that is both pleasant and accommodating. But to bring the squash even more to the forefront, use clean, neutral rice wine vinegar, punctuated with mint and a little black pepper.

YOUNG, FRESH SUMMER SQUASH, SUCH AS ZUCCHINI AND CROOKNECK
   (MULTIPLE KINDS IF POSSIBLE)

SALT AND FRESHLY GROUND BLACK PEPPER

RICE WINE VINEGAR

MINT, TORN OR CUT INTO BITE-SIZE PIECES

Cut off the inedible stem and base of the squash, then shave lengthwise into ribbons on a mandoline or crosswise into discs. Toss sparingly with the salt and black pepper and the vinegar to taste, taking care not to drown out the delicate flavor of the squash. Add the mint and serve immediately. (Squash is mostly water, so if you let it sit, the salt will draw out the liquid from the squash and dilute the flavor of the dressing.)

👉 Rice wine vinegar's sweet, gentle acidity is the perfect seasoning for any shaved vegetable preparation—carrots, fennel, even firm lettuces like endive. The lack of oil gives the salad a clean, direct flavor that highlights the vegetable.

Salted

Salting raw ingredients is one of the most ancient forms of preservation—and of cooking. Sometimes, as with jerky or salt cod, the salt acts as a drying agent, a process called curing. Other times salt is added to a raw ingredient to begin a fermentation process, as with soy sauce, miso, or pickles.

Not all salt processes are long ones. Salting onions draws out their bitter, sulfurous notes, making them more palatable. It softens watery vegetables like radishes or cucumbers, giving them a texture neither raw nor cooked.

The amount of salt, the length of time salted, and the temperature at which an ingredient is held after salting are the main factors in determining the effect of salting. When salt is lightly applied to raw fish, it draws moisture out of the flesh, making it firmer and more dense and concentrating its distinct umami flavors. Gravlax and cold-smoked salmon are good examples of this method.

Gravlax

1 CUP SALT

1 CUP SUGAR

½ CUP MINCED DILL OR FENNEL FRONDS

FINELY GRATED ZEST OF 2 LEMONS

3–5 POUNDS SALMON FILLET, IN ONE PIECE

Combine the salt, sugar, dill or fennel, and lemon zest in a bowl. Spread the mixture evenly over the top of the fish—you might not use all of the salt mixture—then cover the fish with plastic wrap. Put it on a tray with a little lip, to capture the liquid that will be released. Put a flat plate or tray on top of the plastic wrap, and a weight of about a pound on top of that. Leave in the refrigerator for three days.

Remove the weight and discard the herbs. If you are not going to serve the salmon immediately, rewrap it tightly. Otherwise, slice it thinly against the grain, which creates a more elegant texture (gravlax is very flavorful and big chunks would be unpalatable), and serve it at cool room temperature. Removing it from the refrigerator and allowing it to warm slightly before serving brings out the oils in the fish, intensifying the flavor.

Gravlax goes well with cultured cream, like crème fraîche, which adds fat and tang. Baked or griddled bread such as toast or blinis gives relief from the intensely fishy and mellow yeasty flavors. Salty-sour capers and lemon provide contrast, as does a bit of freshly ground black pepper.

Pickled

There are a few ways to pickle foods. One is to simply salt an ingredient, usually cut, and allow it to ferment. This process creates a strong flavor, marked by a sourness that comes from the fermentation process itself rather than the clean, sharp flavor of vinegar. It is one of the most ancient ways to preserve vegetables like cabbage, as with sauerkraut and kimchee. This process can take days or months.

Macerating an ingredient with vinegar and salt is an easier method. This is sometimes referred to as a quick pickle, because the cook is using the fermentation of the vinegar to accelerate the pickling process. The vinegar is usually diluted in water and often balanced with sugar, depending on the level of intensity desired. The vinegar mixture can be added cold to the other ingredients, or heated to cook and soften them slightly and bring out a sweeter, milder taste.

Pickled Mushrooms with Tarragon

Button mushrooms start out hard and chalky; pickling turns them soft and juicy, not raw but also not cooked. Their blandness is a perfect host for the light, fine licorice flavor of tarragon.

1 POUND BUTTON MUSHROOMS

1½ CUPS WHITE WINE VINEGAR

1½ CUPS WATER

2 TABLESPOONS SALT

1 TABLESPOON SUGAR

2 TABLESPOONS CHOPPED FRESH TARRAGON

Clean the mushrooms and put them in a bowl. In a large saucepan, combine the vinegar, water, salt, and sugar and bring to a boil. Stir to combine and dissolve the solids. Add the tarragon, then pour the mixture over the mushrooms. Put a plate or other weight over the mushrooms to keep them submerged. Let them cool to room temperature, stirring occasionally, then refrigerate until needed.

☞ These mushrooms are mellow enough to make a great addition to a crudité plate, or they can be chopped and added to a simple butter sauce to lend acidity and create a more complex flavor.

Fermented

Fermenting is probably not something most people will do at home, although for those who are interested, there are some very good books, like Sandor Katz's *The Art of Fermentation*, and classes and workshops are increasingly available. You can buy many useful ingredients that are already fermented, however.

Soy sauce, sauerkraut, vinegar, wine, beer, and pickles are all examples of fermentations that are commonly used in cooking. From a cook's perspective, the flavors produced by fermentation can add incredible depth and complexity to simple preparations. In much the way that browning meat adds hundreds of aromatic molecules to a dish, fermentations introduce a wide array of facets to your primary ingredients, like a massive new pool of marriage candidates. They can provide a contrast, as with a salty, umami soy glaze punctuating sweet potatoes, or a unifying flavor, the way that the floral sweet-sour-salty facets of white miso bind with tomatoes in a sauce. Fermented flavors can announce themselves loudly or quietly, to reinforce a flavor or to add underlying savory notes.

Fermented Mushrooms

This is a simple variation on the pickled mushroom recipe that illustrates the difference between adding and creating fermented flavors. The vinegar-pickled mushrooms are a bright condiment. These fermented mushrooms add gentle acidity and depth to pastas, stews, and many other dishes. Think of them as an ingredient as much as a condiment.

This process is called lacto-fermentation, which is the same way that sauerkraut and kimchee are made. In an anaerobic environment, lactic acid bacteria like lactobacillus (commonly referred to as probiotics) convert the natural vegetable sugars to lactic acid. The salt preserves the texture of the vegetables and discourages the growth of other, nonproductive bacteria. It sounds complicated, but the recipe is easy, and you can substitute almost any other vegetable for the mushrooms. You can also vary the flavor by adding spices.

4 CUPS FRESH SHIITAKE MUSHROOMS

1 TABLESPOON SEA OR KOSHER SALT (DON'T USE IODIZED, WHICH WILL INHIBIT THE FERMENTATION)

1 QUART WATER

Wipe the mushrooms clean and leave them whole. Dissolve the salt in the water. Pack the mushrooms in a glass jar with a tight-fitting lid. Pour the salted water over the mushrooms until they are completely submerged, then put on the lid. Let them sit at room temperature for two to four days. Taste occasionally. When the mushrooms have a pleasant sourness and complex aromatics, they are done. Refrigerate until needed. Use as a condiment, or to flavor soups and sautés.

☞ Spices are really just non-leafy parts of a plant that have been dried and that are used to flavor food. If you think beyond the supermarket, many flavors can be spices. For example, you could dehydrate these mushrooms, grind them in a spice grinder, then use them like a spice. Slightly salty and deeply umami, the powder would be great for dusting roasted chicken or adding to cooked vegetables.

Soy-Glazed Sweet Potatoes

SERVES 4

This is an incredibly simple recipe with very complex flavors. Sweet potatoes are baked, their flesh is scooped out and gently broken with a fork, and then they're dressed with a mixture of brown butter and dark soy sauce. The meaty, browned facets of brown butter and soy bind together to make a harmonious new flavor.

4 SWEET POTATOES
8 TABLESPOONS BUTTER
4 TABLESPOONS DARK SOY SAUCE

Bake the sweet potatoes at 350 degrees until cooked. Cut them in half, then lightly crush the insides with a fork. In a small saucepan, heat the butter over medium-high heat until it foams and the solids brown. Remove the pan from the heat and stir in the soy sauce. Spoon the mixture over the sweet potatoes.

☞ If you want to brighten this dish, add grated ginger to the butter, or a little lime or lemon zest at the end.

Tomato-Miso Sauce

MAKES ABOUT 4 CUPS

Although we know them mostly as an ingredient in savory food, tomatoes are actually a fruit. They are sweet, and they require salt or acid to create balanced flavors. Cooked into a pasta sauce, tomatoes are often accompanied by green, floral notes like basil, or salty ones like olives and capers. From there it's not a big stretch to consider flavoring a tomato sauce with miso.

Miso is a fermentation of cooked grains (usually rice), salt, and a bacterium called Aspergillus oryzae. As the mixture sits at warm room temperature over a period of months, a complicated reaction occurs between the starches and the sugars that produces savory, meaty flavors with a strong backbone of salt. The several different kinds of miso are distinguished by the different grains on which they are based and the varying amounts of time for which they are aged, resulting in flavors that range from elegant to rustic and earthy.

A white miso—less aged and more subtle than other misos, with fruity notes— matches the sweetness of the tomatoes and provides a subtle floral depth that melds with it in an almost invisible way. A little bit of Parmesan added at the end locks with the miso to create an underlying umami presence.

2 POUNDS RIPE TOMATOES

4 TABLESPOONS WHITE MISO

½ CUP FRESHLY GRATED PARMESAN

SALT

Core and quarter the tomatoes. Put them in a sauce pot with the miso and cook them over low heat until they collapse and are tender through and through, about 20 to 30 minutes. Pulse in a blender with the Parmesan.

Add salt if necessary. Use as you would any tomato sauce—on pasta and so on. It's highly versatile.

☞ This sauce is a great braising medium for lamb, beef, or pork shanks, which are gelatinous and sweet.

{ COOKED }

Heat changes the flavor and texture of ingredients. In some cases, as with potatoes, these changes are necessary to make the ingredient not only appealing but also digestible.

There are many kinds of heat-based cooking processes, but they all fall under two broad categories: soft-cooking processes, in which the ingredients don't brown, and hard-cooking processes, in which they do.

Soft cooking is a good choice when the ingredients are special and distinctive in their original state and you want to preserve their flavor as much as possible. For example, if you have sweet baby turnips, you may want to simmer them in water just long enough to make them tender, retaining and highlighting their sweetness and freshness.

Soft cooking can also be a step in a multiple-step cooking process. For example, you might poach a piece of meat or a vegetable until just tender, then quickly sear it to add another layer of flavor. You might cook onions slowly until tender—"sweat" them—to bring out their sweetness as the base for a soup before adding liquid. Or, as in the Cauliflower with Cumin Seed and Browned Butter (page 70), you might cook cauliflower in salted water to soften it and then brown it in butter.

Hard-cooking techniques are good for adding depth and richness. They can help bring out the complexity of flavor-dense foods like meats, or they can add browning flavors to ingredients in need of a boost—for example by roasting those turnips at high heat if they happen to be not young and crisp but past their prime.

Generally, hard-cooking processes add flavors not present in the ingredients themselves.

Water

Water makes up most of the world around us, and it is a major component of most ingredients. Water also is added into many dishes, not only as water itself but in the form of broth, wine, juices, and other liquids.

Along with mastering seasoning, controlling water content is one of the most important things a cook can learn. Adding or subtracting liquids changes both the texture and the flavor of the food you are making. Too much water in a soup or a sauce makes it thin and bland. Too little water and the flavor becomes dense and muddy. In perfume making as in cooking, just the right amount of dilution is key. The physical form of the dish affects both the aromas and the flavors. A thick sauce stays on the tongue, giving it more power and presence. A thin broth is evanescent, fleeting, and can be quickly replaced by a more substantial bite.

The process starts at the ingredient level, by either lessening the water or adding more. The salted cucumber is an example of how salting can draw out liquid, creating a denser texture and more intense taste. Soaking grains like bulgur or farro in water overnight hydrates the grains, allowing them to cook more quickly and evenly, especially when steamed (which retains the textural integrity of the individual grains). Water also affects the cooking process, as we'll explore in this chapter. Too much moisture on the outside of a piece of meat will prevent it from browning. Too little moisture in a stew and it will burn.

Temperature

Serving temperature has a big impact on how flavor is released to the mouth and nose. Hot food is generally more aromatic than cold food, because heat volatilizes the aroma molecules more rapidly. Sometimes the muting effect of lower temperature is just what we want: a pickle is best enjoyed cold, because heat would make the acidity overpowering and unpleasant. A chicken broth, on the other hand, wants to be warm, because cold it will be sticky, gelatinous, and dull-flavored.

Temperature, along with water, defines most cooking processes. Poaching a chicken breast—gently cooking it in hot liquid—preserves its moistness by allowing a slow, steady application of heat; raise the temperature too much and you end up with flesh that is tough and dry on the outside and undercooked within. Conversely, too low a temperature when you are attempting to sear a steak releases moisture on its surface and impedes browning; the meat steams rather than sears.

{ SOFT COOKING }

Sweating

Sweating is the soft-cooking cousin to sautéing, which we'll get to below. It involves cooking an ingredient in a small amount of fat but at low heat—usually covered or partly covered—to bring out its juices. An onion cooked in this way has a simple sweetness. Onions that are sweated before being added to soup have a different taste—softer, sweeter—from onions that are dropped in raw.

To sweat onions, thinly slice or chop them and put them in a pan with a little butter or oil and some salt. The fat keeps the onions from sticking to the pan, and the salt draws out the water in them as they cook. Turn the burner on low and cover the pan. The low heat allows the onions to cook without browning, and the lid captures the moisture and keeps it in the pan. The liquid that comes out of the onions contains vegetable sugars, and the heat concentrates the sweetness. Sweated onions should be completely soft, not at all browned, and very sweet, with none of the acrid notes of raw onion.

Boiling

Boiling is cooking ingredients immersed in very hot—bubbling—liquid. Its close relative, simmering, occurs just a temperature-step down, with the liquid barely bubbling (usually after it has reached a full boil first).

Simmering or boiling ingredients in salted water or a flavorful liquid can be the start of a dish that uses both the liquid and the ingredients that get added to it: the ingredients both absorb the flavors in the liquid and alter it. This is a good technique for creating composite, layered flavors, like a curried squash soup that uses chicken stock and coconut milk.

Boiling ingredients in salted water is also a good way to focus the flavor of an ingredient, making it taste only of itself. This is a nice technique for cooking vegetables that you want to use in a vegetable salad, for example, when you want each vegetable to retain its bright color and distinct character. Cooking vegetables in salted water, rather than steaming them, seasons them all the way through. As a general rule, about 3 to 3.5 percent salt to water by weight—about the salt level of seawater—is a good ratio for cooking vegetables. That's a little less than 2 tablespoons of salt per quart of water.

Boiled Scrambled Eggs

SERVES 2

Boiling eggs can lead in several directions. There are soft-boiled eggs, with a runny yolk, and hard-boiled eggs, perfect for deviling or for slicing onto a salad. This recipe is for boiled eggs that have been beaten. The intense heat of the water causes the air bubbles in the eggs to expand and simultaneously sets the proteins, making the eggs light and delicate, but fully cooked. A surprisingly fast, easy, and delicious recipe.

2 VERY FRESH EGGS

SALT

Crack the eggs in a bowls and whisk vigorously with a fork for 30 seconds. You are both emulsifying the fat of the yolks with the rest of the eggs and incorporating air bubbles.

Set a fine-mesh strainer basket in the sink. Fill a medium pot with water to a depth of at least 4 inches, then bring to a boil. Stir the water to create a whirlpool effect—so that the eggs don't fall to the bottom and stick—and then pour in the eggs. Cover, turn down the heat to medium, and count to twenty. Uncover. The eggs should be floating on the surface in ribbons. Gently pour into the strainer, let drain for a minute, scoop into bowls, and serve.

☞ These eggs taste even better when enriched with some kind of fat. In the spring and summer, olive oil and a flurry of herbs and edible flowers. In winter, a small amount of milk infused with rosemary, combined with half as much fruity olive oil. The eggs are also great with more traditional accompaniments, from crème fraîche and chives to hash browns and grilled onions.

Steaming

Steaming is cooking ingredients over, not in, boiling water, in a contained environment. It allows fish, meat, and vegetables to cook in a moist environment without being immersed in water, which can dilute flavor and degrade texture, especially of delicate ingredients. Steaming at low temperatures can be a gentle cooking process, especially good for applying heat to fish and greens. Steam happens across a range of temperatures, so the heat can be adjusted to the ingredient: a piece of fish wants a lower temperature, to allow the proteins to coagulate softly; grains, a higher temperature, to cook them.

Steamed Halibut with Bok Choy and Soy-Lemon Sauce

SERVES 4

This recipe works with any white-fleshed fish. Steaming preserves its delicate texture and flavor and keeps it moist and tender. Because there are no additional flavors added from a browning process, the fish must be very fresh.

1 POUND FILLETS OF HALIBUT OR OTHER WHITE-FLESHED FISH,
   ½ TO ¾ INCH THICK

SALT

½ POUND BOK CHOY, SEPARATED INTO LEAVES

2 TABLESPOONS DARK SOY SAUCE

2 TABLESPOONS FRESHLY SQUEEZED LEMON JUICE

½ TEASPOON CHILI FLAKES

2 TABLESPOONS VEGETABLE OIL

Put a bamboo or other steamer in a pot with at least an inch of simmering water below. Season the fish with salt, place it on the steamer, and cover. After 2 minutes add the bok choy and cook for 2 minutes more, or until the fish is done.

While the fish is cooking, combine the remaining ingredients. When the fish is cooked and the bok choy is tender, divide them among serving plates, and drizzle the sauce over them.

☞ In this recipe the vegetable oil creates mouthfeel and connects the soy and lemon without imposing any flavors of its own. Sometimes you want a fat to bring flavor, and other times, as here, you want it simply to carry the flavors around it.

Poaching

Poaching is cooking immersed in liquid, at a lower temperature than boiling or simmering. It's a technique usually used for meat and fish, when a moist and tender result is desired. As opposed to steaming, however, poaching does add flavor from the cooking liquid, and the poached ingredient can flavor the poaching liquid in turn. Poaching chicken in a seasoned chicken stock both flavors the meat and fortifies the broth, for example. Poaching is a good choice for a delicate preparation where you want to amplify flavors or add aromatics.

Chicken Breasts Poached in Ginger Broth

SERVES 2-4

Boneless chicken breast is always challenging to cook because it has so little fat and connective tissue that it quickly turns dry and mealy. Poaching allows the chicken breast to cook slowly in a moist environment and imbues it with more flavor than it naturally has. You could serve it just this way, accompanied with a little rice (simmer it in the broth as well). It is also perfect to use in salads, sandwiches, taco fillings, or any other way.

1 QUART CHICKEN STOCK

4 SLICES GINGER

1 TEASPOON WHOLE BLACK PEPPERCORNS

ZEST OF 1 LEMON, IN STRIPS

SALT

1 POUND BONELESS, SKINLESS CHICKEN BREASTS

Bring the chicken stock to a boil. Reduce to a simmer and add the ginger, peppercorns, lemon zest, and salt to taste. Simmer the stock for 5 minutes and then add the chicken breasts. Turn off the heat, cover, and let the chicken breasts sit in the pot until they are cooked through, 15 to 20 minutes. Strain the enriched broth and reserve it for another purpose.

 This is also a great way to poach fish. Replace the chicken stock with vegetable stock and increase the aromatics, perhaps adding lemongrass or chili.

Smoking

Smoking is an ancient method of flavoring and preserving foods, especially meats. Even more than grilling, it gives foods a primal and deeply satisfying flavor. Smoke adds umami and intensity, locking with the flavor of the ingredients and amplifying it.

There are two kinds of smoking. In hot smoking—for instance barbecued ribs—the heat is high enough to cook the ingredients. In cold smoking—for instance bacon or smoked salmon—the temperature is low, and the ingredients are not cooked, but just infused with the smoke.

Smoking doesn't require fancy equipment. You can buy smokers specifically designed for hot and cold smoking. You can also build your own. For hot smoking, all you really need are a metal container with a lid, some wood, charcoal, or propane for heat, and some smoldering wood chips in an outdoor or at least heavily ventilated space. You can hot smoke on a regular gas or charcoal grill, or even on a purchased or improvised stovetop smoker. Cold smokers are designed to allow ingredients to be held in a separate chamber from the burning wood or coal; the smoke they generate is pumped through to flavor the food without heating it. In the restaurant we improvise a cold smoker by placing a piece of wood (usually local cherry or apricot) over a flame until it starts to smolder, then putting it in a metal pan next to the ingredient we want to smoke and covering it with foil and letting it sit at room temperature (we put the food on a piece of ice if we need to keep it cold). The wood smokes but doesn't add much heat. If we need more smoke we remove the piece of wood, reignite it, then put it back, but when it burns down too far we discard it, to avoid a bitter charcoal flavor. You can find instructions for building and using both kinds of smokers online.

But smoky tastes don't have to be something you create on your own. Ingredients like bacon and ham hocks bring the smoke with them, as do smoked paprika and dried chipotle chilies, or oil that has been smoked, which carries only the flavor of smoke in a neutral base.

Smoked Oil

One great way to capture the flavor of smoke is by cold-smoking oil. The smoked oil can be used in sauces and vinaigrettes, drizzled over steamed vegetables or fish, or added to any other preparation that could benefit from the umami kick of smoking. It's like vegetarian bacon, the smoke without the pork.

PURE OLIVE OIL (OR SUBSTITUTE ANOTHER KIND OF OIL IF YOU WISH)

Place the oil in a bowl. Put the bowl, uncovered, in a cold smoker and smoke for 20 to 30 minutes, until the oil is flavorful. You may store the oil in a tightly sealed jar in a cool place for up to a month.

Smoky Beans

SERVES 6-8 AS A SIDE DISH

In this recipe the smoked oil infuses meaty cranberry beans that are baked to preserve their texture. The concentrated cooking creates a "bean gravy," made complex and beguiling by the smoke flavor, which evokes beans cooked with ham hocks or bacon.

1 POUND DRIED CRANBERRY BEANS

1 MEDIUM ONION, DICED

1 LARGE CARROT, PEELED AND DICED

4 TABLESPOONS SMOKED OIL (SEE ABOVE)

4 TABLESPOONS CHOPPED GARLIC

1 TEASPOON CHILI FLAKES

2 QUARTS WATER

SALT

Rinse the beans and place them in a large bowl or pot and cover them with water to a depth of at least a couple of inches. Let them soak overnight, then drain.

In a medium skillet, sauté the onion and carrot in 2 tablespoons of the Smoked Oil over medium heat until they are lightly browned. Add the garlic and continue to cook for another minute or two, until the mixture is aromatic. Stir in the chili flakes and put the mixture in a Dutch oven or other large, heavy ovenproof pot with a tightly fitted cover.

Add the beans, the water, and the remaining Smoked Oil. Cover and bake at 350 degrees for 2½ to 3 hours, opening the pot once or twice to stir the beans and to check that there is enough liquid; if not, add a little more water. When finished, the beans should be tender and there should still be some "bean gravy" in the pot. Add salt to taste, and a little more Smoked Oil if needed.

{ HARD COOKING }

The Maillard reaction—commonly called "browning"—is a chemical reaction between amino acids and sugars that happens in situations with high heat and low moisture. This reaction creates hundreds of flavor compounds, which keep evolving as the process continues. They contribute an incredible array of savory umami flavors to food, which is why grilled and roasted foods are so delicious. They also create a host of aromatic compounds that make browned food smell mouthwatering even before you taste it.

Sautéing

Sautéing refers to the process of cooking ingredients quickly at high heat with a little bit of fat in a low, wide pan that allows released liquids to evaporate quickly. Sautéing is a low-moisture cooking process that often leads to browning.

An onion that is sautéed will lose its bite, as its flavor becomes sweeter and more gentle. As it browns, it will become more complex, taking on a little bitterness and some oxidative notes, like you find in sherry. Young carrots that are boiled are sweet and simple. Older carrots that are sautéed develop secondary aromatics that give them a meaty flavor.

Sautéing is best used when the time needed to cook an ingredient to desired doneness is relatively short. For ingredients that need longer cooking times, pan-frying and roasting work better. But sautéing is also a step in many dishes, to develop complex flavors in vegetables that will form the base of a soup or stew, for example, or before liquid is added to create a braise, as we'll describe below.

Sautéed Spinach

SERVES 4

Sautéing spinach very quickly at high heat allows it to wilt, giving up its moisture and concentrating its flavor. Spinach is very delicate, with high water content, so take the following steps to ensure that it sautés rather than stews, which will ruin its texture and dilute its flavor: Dry the spinach thoroughly before cooking. Use a wide, open pan that will facilitate evaporation, and don't add too much spinach to the pan at once. If you have a lot of spinach, cook it in batches.

2 TABLESPOONS BUTTER

1 POUND SPINACH, WASHED AND THOROUGHLY DRIED

SALT

In a wide, heavy sauté pan, heat the butter over high heat to bubbling. Add the spinach and some salt. Stir the spinach a few times until it is just cooked; it should take less than 30 seconds. Transfer it to a serving bowl.

☞ This method works with any tender green and any kind of fat—for example, fava leaves and sesame oil.

Pan-Frying

Pan-frying is cooking food over high heat and partially immersed in fat. Like sautéing, it uses a wide pan. The depth of fat can range from as little as a quarter inch to halfway up the side of the ingredient. Pan-frying is a good choice when an ingredient will take so long to sauté to doneness that it will lose too much moisture. The greater proportion of fat is a vehicle for high heat that both penetrates the ingredient, cooking it through, and browns it. It's also a great way to achieve a crunchy exterior or crust, as with starch-coated meat or fish.

Pan-Fried Chicken Thighs
SERVES 4

This is a classic take on fried chicken that gets a boost from the addition of smoked paprika. Controlling the heat is crucial. Too hot and the outside will burn before the inside cooks. Too low and the skin will become soggy and limp, the meat oily.

You can use chicken breasts in this recipe, but it's not recommended, as their lack of fat will make them dry out in the intense heat of pan-frying. It's also best to leave the chicken on the bone. The bone will flavor the meat (for the same reason that bones flavor stock) and keep it moist.

Besides the paprika, two kinds of pepper are added to the flour that coats the chicken. The black pepper gives a bold spiciness that hits the taste buds immediately and quickly recedes, and the cayenne gives a subtle, fruity spice that adds complexity.

SALT
8 CHICKEN THIGHS
2 CUPS BUTTERMILK
2 CUPS FLOUR
1 TABLESPOON SMOKED PAPRIKA
2 TEASPOONS CAYENNE
FRESHLY GROUND BLACK PEPPER
VEGETABLE OIL FOR FRYING

Salt the chicken thighs and let them sit uncovered in the refrigerator for at least 2 hours, to air-dry the skin. Put them in a bowl, then mix the buttermilk with ½ tablespoon salt and pour the mixture over the chicken. Cover the bowl with plastic wrap and place it in the refrigerator overnight.

Drain the chicken well. In a mixing bowl, combine the flour and spices. Dredge the chicken in the flour mixture and shake off any excess. Place the floured chicken pieces on a rack and let stand at room temperature for 30 minutes, then dredge in the flour again.

Heat about 1 inch of vegetable oil in a cast-iron skillet over a medium-high flame. (A cast-iron skillet is best: it is slow to heat up, but once it does, the thickness and density of the metal help it retain the heat, with the temperature of the oil kept more consistent and the cooking process regulated so that the chicken cooks more evenly.) Test the temperature of the oil by adding a pinch of flour. If it immediately bubbles and sizzles, the oil is ready. Add the chicken, skin side down. Don't crowd the pan, or the chicken will steam rather than brown. If you have more chicken than will fit comfortably, use two skillets or fry in batches. Cook the chicken, turning frequently, until the skin is a deep golden brown and the meat is cooked, about 20 minutes. Remove the pieces to a paper towel to drain.

You can serve the fried chicken immediately and enjoy it hot and juicy, the skin crackling crisp. Or you can refrigerate it and eat it cold the next day. The juices will recede into the meat, and the cold will leave the flesh more dense, but still moist. The crust will soften, and the flavors will meld. Chicken served this way is another dish entirely—comfort food, soft and flavorful, a unique pleasure.

 If you have time, allow the chicken to cure with the salt overnight.

Deep-Frying

Deep-frying is the same as pan-frying, except that the ingredient is fully submerged. Frying is a good technique for burning off all an ingredient's surface moisture when a crisp exterior is desired, as with potato chips or fries. It also delivers a more consistent heat than does pan-frying, which requires constant turning to even out the browning process. That makes deep-frying good for fritters and other preparations for which it's desirable for all the exterior to cook quickly at once.

In *The Physiology of Taste*, Brillat-Savarin captures the appeal of deep-frying poetically:

> Fried things are highly popular at any celebration: they add a piquant variety to the menu; they are nice to look at, possess all of their original flavor, and can be eaten with the fingers. . . . Frying also . . . comes to [the aid of cooks] in emergencies; for it takes no longer to fry a four-pound carp than it does to boil an egg. The whole secret of good frying comes from the surprise; for such is called the action of the boiling liquid which chars or browns, at the very instant of immersion, the outside surfaces of whatever is being fried. By means of this surprise, a kind of glove is formed, which contains the body of food, keeps the grease from penetrating, and concentrates the inner juices, which themselves undergo an interior cooking which gives to the food all the flavor it is capable of producing. In order to assure that the surprise will occur, the burning liquid must be hot enough to make its action rapid and instantaneous; but it cannot arrive at this point until it has been exposed for a considerable time to a high and lively fire.

Fries

Good fries should be crisp outside and fluffy and steamy inside, well-seasoned with salt, and tasting of potato, not fryer oil. To achieve the best result, use starchy potatoes like russets, which will attain a lightness to the interior that nicely balances the rich, crunchy exterior. Waxy potatoes like creamers will be dense inside as they cook, which will not create the desired textural contrast with the crisp outside.

The proportion of outside to inside is also crucial. Skinny fries will have more surface area exposed to the oil, making for a greater proportion of crunchy exterior. Larger fries, with more internal mass, will generate more steam, which makes the fries limp. And of course, keeping the fries at the same size allows them all to cook at the same rate. Quarter-inch bâtons are a happy medium.

One other thing: To make fries properly, you need to fry them twice—once at a lower temperature to make them tender and draw out the excess moisture, and once again at a higher temperature to crisp the outside. You can do the initial frying and hold the fries for several hours or overnight before performing the final crisping.

VEGETABLE OIL

RUSSET POTATOES CUT INTO ¼-INCH BÂTONS

SALT

Pour the vegetable oil to a depth of at least 4 inches in a fryer or heavy pot (cast iron is good) outfitted with a thermometer, and place it over high heat. When the oil reaches 325 degrees, add the potatoes, being careful not to crowd the pan. (If you have more potatoes than you can comfortably fit, cook them in batches.) Fry until the potatoes are very tender, about 8 minutes, maintaining the temperature. If they're not cooking in a fry basket, remove them with a slotted spoon to a paper towel–lined tray or baking pan.

Just before serving, heat the oil to 375 degrees. Return the fries to the pan and cook until they are golden brown and crisp, about 3 to 4 minutes more. Drain well, toss with salt, and serve piping hot.

☞ Try adding some fresh rosemary and sage leaves at the very end of cooking. They will crisp and add complex aromatics.

Searing

To sear food is to brown it quickly in a moderate amount of fat over very high heat. This method is sometimes used for foods that must brown quickly without allowing the insides to overcook. Scallops and liver are two ingredients that benefit from this method.

Like many other methods, searing can be one step in a longer cooking process. It can be a prelude to braising. It can brown bones that will be the base of stocks and sauces. It also works well for secondary cooking—for example, with vegetables that have been steamed or boiled until they are almost tender and just need to be quickly browned. Broccoli that has already been boiled, drained, and tossed in olive oil is excellent finished this way, with the browning on a hot grill adding delicious smoky flavors.

Seared Scallops with Butter and Lemon

Scallops are one of the most delicious and easy-to-use sea animals. They are a white, tender muscle that grows in a shell, and while it is possible to find them in their natural state, usually they are sold already shucked. There are tiny bay scallops and large sea scallops. Both kinds are sweet, gently oceanic, and accommodating of many flavor partners. Try to find scallops that have not been treated with preservatives, as those that have been tend to retain water and are harder to brown.

Scallops require only a very quick searing over high heat, but it can be tricky to get them to brown rather than steam, because of their high moisture content and the brief cooking time they need. Moreover, they are delicious with butter, but butter burns at the high heat needed for browning. The solution, here as elsewhere, is to use multiple fats. Here, we start by searing the scallops in vegetable oil, then add the butter after the scallops are turned. A little lemon juice at the end, and that's it. Simple and quick, but very delicious. Have all the ingredients and accompaniments ready before cooking the scallops, because they take only a minute, and they should be served the second they're finished.

VEGETABLE OIL

SEA SCALLOPS

SALT

BUTTER

FRESHLY SQUEEZED LEMON JUICE

Place a sauté pan over very high heat and add the vegetable oil. While it heats, pat the scallops dry—any excess moisture will impede browning— and season them well with salt. (The butter and lemon will tamp down the seasoning.)

When the oil is almost smoking, add the scallops to the pan. (If you are using large sea scallops, put the nicest side down.) Cook for 45 to 60 seconds, until the undersides are deeply browned.

Turn the scallops over and lower the heat to medium. After 15 seconds, add the butter, which will start to bubble and brown instantly. Move the fat around with a spoon so it doesn't burn; as soon as it is browned, add the lemon juice. Cut the heat and spoon the lemon butter over the scallops a few times to glaze them. Plate the scallops and drizzle them with the butter sauce.

☞ This is a base recipe that can be built in many ways. Whisk chili paste, miso, or soy sauce into the lemon butter, or add a variety of citrus, as in the sauce for Asparagus with Citrus Sauce (page 135).

Grilling

There is nothing quite like cooking over an open flame. When people first discovered cooking over a fire, they learned about fragrance and flavor simultaneously, and grilling still taps into that primordial experience. Simply grilled meats and vegetables are perfect by themselves, often requiring only a simple accompaniment to make a meal. But a combination of smoke, herbs, and spices can also be enchanting, creating flavors that layer with the umami tastes of deep browning and take grilling a long way past steak.

But first, there's steak.

Grilled Steak

STEAK, ANY CUT

SALT

VEGETABLE OR OLIVE OIL

FRESHLY GROUND BLACK PEPPER

Prepare a grill. When the coals are ready or the gas is fired up, season the steak with salt. (Don't grind on the black pepper until the end: it will burn, and it begins losing its aromatic notes as soon as it is ground.)

Lightly oil the steak and place it on a hot part of the grill for 30 to 60 seconds. Turn it over and cook it for another 30 to 60 seconds. Remove it to a plate for 30 seconds. (This seems counterintuitive, but it's a better way to treat primal muscles like steak. Muscle is dense, and heat penetrates it very slowly, with the heat transferring from one cell to another. That means that when you apply heat to the outside, it takes a while for that heat to get to the center. Briefly removing the steak from the heat allows the surface heat to slowly work its way inward. This keeps the internal temperature lower during cooking, which leads to a juicier, more tender piece of meat.) Keep rotating the meat from grill to plate in this way until it's just about as done as you want it. Test it by pushing with your finger to feel the resistance. The firmer it gets, the more cooked it is—develop a feel for just how you like it.

Right before serving, return the steak to the grill, turning it frequently, until the interior is hot and the surface is nicely browned. Remove the steak to a plate and season with black pepper.

☞ This is a great method to use when you've got a lot on your plate, so to speak—multiple dishes to prepare for your guests, say. An hour or so in advance, take the steak almost all the way to done to your likeness this way, then pull it off the heat. That will give you time to pull the rest of the dinner together, and then when you're ready, you'll need only another minute or two to finish the steaks so that they arrive at the table hot from the grill and perfectly cooked.

Baking and Roasting

Baking and roasting are basically the same process—cooking foods with dry heat in an oven. Typically, baking refers to breads and pastries, roasting to meat and vegetables. Roasting also tends to be used to refer to deeply browned foods, even though the Maillard reaction is also a key component in the complex flavors of bread and many kinds of pastries.

Whatever we call them, baking and roasting can happen either at lower temperatures, which create less browning, or higher temperatures, which create more. As with other methods, the choice of process and temperature derives from both the qualities of the ingredients themselves and the desired flavors, textures, and aromas. Large cuts of meat like ham and brisket benefit from long, low cooking, often covered to preserve moisture. This allows their tough connective tissue and collagen to break down into a soft, luscious texture.

Smaller cuts like rack of lamb require searing followed by moderate heat, as they also contain some tough bits that need time to become tender and delicious. High heat is good for birds and smaller cuts of meat like pork chops that can become tough and dry with long cooking. High heat is also good for bread, because it forces the water and air inside the dough to expand quickly and create a light crumb.

The *Li Chi*, the only Chinese book of recipes and rituals that survives from the second century BC, embodies an intense thoughtfulness about exercising the control necessary to master the outcome of a dish, including an often elaborate sequence of cooking processes. For example, as the pièce de résistance of eight delicacies considered appropriate for the aged on ceremonial occasions, there is what has to be the most elaborately cooked pig ever. Reay Tannahill describes the process in *Food in History*:

> The stuffed piglet was wrapped in a jacket of straw and reeds coated with wet clay, then roasted until the clay dried, by which time the juices had been sealed in and the skin had softened. The clay was discarded, and the skin stripped off and pounded to a paste with rice flour and a little liquid. Recoated with this paste, the piglet was deep-fried until it was cooked to an appetizing golden brown. Then came the final stage, when the meat was sliced, placed on a bed of herbs and steamed gently for three days and three nights. When it emerged, exquisitely tender and aromatic, it was served with (something of a jarring note, this) pickled meats and vinegar.

Roasted Root Vegetables

Root vegetables like celery root, parsnips, carrots, and onions are hardy and deeply fla-
vored, and they respond well to roasting. The high heat draws the sugars to the surface
and caramelizes them, creating complex flavors that play against the sweetness of the
vegetables.

The process is so simple that it doesn't require a formal recipe. Cut the vegetables into chunks of roughly the same size; 1-inch pieces work well, allowing a good balance between browned exterior and tender, sweet interior. Toss them with olive oil and salt and spread them out in a single layer on a roasting pan. Roast at 400 degrees, stirring occasionally, until the vegetables are browned and tender.

☞ The variations on this basic technique are infinite. The choice and proportion of vegetables affect the flavor. A garlic head, cut in half, will perfume the vegetables as they cook, as will hearty herbs like rosemary and sage, and spices.

Toasted Bread and Onion Soup

SERVES 6-8

Bread can be an ingredient as well as a food in itself. It develops complexity from both the fermentation of yeasts interacting with flour and the Maillard reaction that happens when it bakes. Both processes contribute to its alluring aroma as well.

Here bread is used as an ingredient in a variation on a baked onion soup: caramelized onions are simmered with chicken stock and then mixed at the end with flavorful bread. The recipe is much easier than traditional versions of onion soup, but the flavor is multilayered: complex browned facets of the bread and onion bind with the sweaty, barnyard notes of aged Gruyère; the garlic adds a fresh allium flavor that locks with the onions, as does the sherry; and the thyme adds a sweet, resinous green note.

4 TABLESPOONS BUTTER

4 MEDIUM ONIONS, THINLY SLICED

SALT

2 CLOVES GARLIC, MINCED

6 CUPS CHICKEN STOCK

2 TEASPOONS FRESH THYME

WHOLE WHEAT BREAD, PREFERABLY LEVAIN, CUT INTO ½-INCH CUBES
  TO MAKE 4 CUPS

2 TABLESPOONS SHERRY

GRATED GRUYÈRE

FRESHLY GROUND BLACK PEPPER

Melt the butter in a heavy saucepan, add the onions and a little salt, cover, and cook over medium heat, stirring occasionally, until the onions are tender, 10 to 12 minutes. Remove the lid, raise the heat to medium-high, and cook, stirring often, until the liquid in the pan is gone and the onions are thoroughly brown.

Add the garlic and cook for another 1 or 2 minutes, until it is aromatic, then add the chicken stock and thyme and bring the stock to a boil. Lower the heat and simmer for 20 minutes.

Stir in the bread and cook for 1 minute more, until the bread is softened. Add the sherry. Season with salt, ladle into bowls, and top generously with the Gruyère and black pepper.

☞ If the onions aren't that sweet, add a pinch of sugar.

Braising and Stewing

Braising and stewing mean basically the same thing: cooking either meat or vegetables over low heat in a moist environment until they are tender. Typically stewing refers to this process as it happens on the stovetop, and braising to the oven-bound version of it. While in this sense they are essentially soft-cooking processes, they tend to begin with hard-cooking processes like sautéing and searing, to introduce deep, roasted flavors that will incorporate themselves into the liquid as the dish cooks. Braising and stewing require ingredients that can stand up to long cooking—a mature hen or even an old rooster (as opposed to a young pullet) and wild mushrooms like morels, which need to be fully cooked.

For searing and braising, heavy clay or enameled cast-iron pots work best, because they hold heat well and distribute it evenly, making it less likely that food will burn. Stainless-steel pots and pans are nonreactive, and will not change the flavor of the ingredients as they cook, especially when acidity is introduced.

Oxtail Braised in Marrow Stock

SERVES 8

Oxtail is one of the most-used muscles in the cow, and therefore one of the toughest—and, when properly cooked, the most delicious. This recipe uses the rich sweetness of marrow bones as the base for tender, meaty stewed oxtail. This stew is versatile and can be used with a variety of accompaniments. When the oxtail is tender, add whatever vegetables you want to the pot—potatoes, leeks, and spinach all go well with it—and simmer until they are tender. Or serve it with grains or pasta.

2 POUNDS SPLIT MARROW BONES

6 POUNDS OXTAIL, IN SEGMENTS

SALT AND FRESHLY GROUND BLACK PEPPER

VEGETABLE OIL

Place the bones in a large, heavy pot with water to cover. Bring the water to a boil, then lower the heat and simmer for 20 minutes. Remove the bones and reserve the stock. When the bones are cool enough to handle, remove the marrow, then blend it back into the stock.

Season the oxtail well with salt and pepper. Add vegetable oil to film a Dutch oven or other large, heavy pot and place it over high heat. When the oil is very hot, add the oxtail and sear on all sides. When it is browned, add the marrow stock to cover, and water if necessary, and bring it to a simmer, scraping loose the browned bits with a wooden spoon. Lower the heat, partially cover the pot, and simmer until the oxtail is tender, about 2 hours.

 Root vegetables cooked in marrow stock, served with grains, make a hearty, vegetable-centric main course.

Orecchiette with Stewed Broccoli, Olive Oil, and Parmesan

SERVES 4

Cooks learn early on not to overcook vegetables, but sometimes overcooking can be delicious. Broccoli, for example, can be cooked until very soft, and then mashed with a fork like potatoes—the fiber breaks down into a soft and creamy mass.

In this recipe, chopping broccoli into small pieces and stewing them slowly and completely brings out their sweetness. The broccoli melts into the sauce and creates a lock with the lemon and olive oil to make a merged, delicious whole. Orecchiette, which means "small ears" in Italian, is a small, irregularly round pasta with a depression in the center that is a perfect vehicle for the sauce. Thin slices of raw, crunchy broccoli stem add a welcome contrasting freshness and texture.

1 HEAD BROCCOLI

4 TABLESPOONS FRUITY OLIVE OIL

1 CLOVE GARLIC, THINLY SLICED

½ TEASPOON CHILI FLAKES

1 CUP WATER

12 OUNCES ORECCHIETTE

SALT AND FRESHLY GROUND BLACK PEPPER

2 TABLESPOONS FRESHLY SQUEEZED LEMON JUICE

LEMON ZEST

FRESHLY GRATED PARMESAN

Separate the broccoli stems from the florets. Chop the florets small. Peel the stems and shave thinly.

Heat 1 tablespoon of the olive oil in a large sauté pan over medium-low heat. Add the garlic and cook for a couple of minutes, until aromatic. Add the broccoli florets, chili flakes, and 1 cup water, bring to a simmer, then cover and cook on low heat, stirring occasionally, until the broccoli is very tender and starting to break down, about 10 to 15 minutes. Add the remaining olive oil and remove from the heat.

Bring a large pot of water to a boil and salt lightly. Add the orecchiette and cook until the pasta has just a bit of chew left in it, about 9 to 10 minutes. Drain, reserving a cup of the cooking liquid. Add the pasta to the broccoli sauce, mix well, and bring to a simmer, then cook together for 2 minutes. Add some of the pasta cooking water if the mixture seems dry, but keep it saucy not brothy—you want the thickened liquid to stick to the pasta. Remove from the heat and season with salt, pepper, and lemon juice to taste.

Divide the pasta among serving plates. Sprinkle with the shaved broccoli stem, and then lemon zest, and grind black pepper over the top. The zest will pull up the lemon aromatics, and the pepper will punctuate the sweet and floral notes. Sprinkle liberally with Parmesan.

☞ In the Apulia region of Southern Italy, orecchiette are traditionally served with rapini, the bitter green we also know as broccoli rabe, but often broccoli is substituted. Orecchiette may be served with sausage—a tasty addition of spice, umami, and fat.

Eight

THE SEVEN DIALS

*If upon a white canvas I set down some sensations of blue,
of green, of red, each new stroke diminishes the importance of
the preceding ones. . . . To do this I must organize my ideas;
the relationships between the tones must be such that it will
sustain and not destroy them. . . . From the relationship I have
found in all the tones there must result a living harmony of
colors, a harmony analogous to that of a musical composition.*

HENRI MATISSE, *NOTES OF A PAINTER*

You've selected a promising combination of ingredients. You've found a direction in seasoning that bridges their differences or creates interesting contrasts, fanning their nuances into a transformative whole you experience as deliciousness. Now you still need to fine-tune the way the dish will taste in the mouth, not only amplifying its contrasts or taming them if they've gotten too wild, but also adjusting for what the dish will be served with, the temperature it will be served at, and any number of other factors.

With food, there are basically seven different kinds of adjustments you can make to balance what you have created: salt, sweet, sour, bitter, umami, fat, and heat. We call them the seven dials. They do not create flavor, but they fine-tune it in a magical way.

The dials are aspects of taste that can be turned up or down, like the knobs on a stereo, to adjust flavor, in concert with the most important tool of all: your own sense of taste and smell. We think of them as a toolkit, because they can be used singly or in concert, and they serve the aromatic ingredients in the dish; they don't replace them.

Especially if you watch a lot of cooking shows, you may be used to thinking of some of these tools—sweet, sour, salt, and bitter—as flavors themselves. Indeed, we often say, "This tastes sour." But sour is not a flavor per se. As we've noted, a lemon does not have the same flavor as a lime, although both are sour. Radicchio and coffee are both bitter, and honey and sugar are both sweet. We can't even say that all salts taste the same. These four basic tastes, important as they are, have dominated the vocabulary of food for too long, undermining our ability to describe flavor with specificity, and to imagine and create it as well. If we focus on them exclusively, we never transcend the static zone of taste and enter the dynamic realm of flavor. But as tools to shape, accentuate, and modulate flavor, they are invaluable.

To these basic four, we add three more. Umami, often called the fifth taste, occurs in the mouth and can best be understood as savoriness. We know from experience that the presence of fat affects flavor, and recent research has discovered taste buds or receptors for fat in the mouth; we consider it here as a sixth taste. And finally, there is "hot," a characteristic of ingredients as diverse as black pepper, chili peppers, horseradish, and mustard. As with the basic four, these are not flavors in themselves; they are *qualities* of ingredients that have their own distinctive flavors, any of which may be key components of the dish you are making. Those qualities are useful when it comes to shaping and balancing—adjusting—the overall experience (flavor) of a dish.

Sometimes you need to compensate for a deficit in an ingredient—a tomato that wasn't as ripe as you would have liked, rosemary that leans too medicinal, or the bland, sugary Golden Delicious apples you had to settle for when you couldn't find sweet-tart Pink Ladies. Sometimes you are adjusting for a lock between ingredients that throws off the balance of a dish, tipping it too far in one direction or flattening it out. Just as they do in perfume, blood orange and jasmine, each sweet in itself, can create a hypersweetness when they lock; orange and tomato, cooked together, can have a similar effect. In that case, a bit of vinegar, a higher level of salt, or the injection of a little heat into the dish can make the combination less sweet and more dynamic and interesting. Sometimes the cooking process throws off a dish—the sauté became so hot that it burned, and you need to add a sweet or bland element to work with the bitter flavors. Or perhaps you buried the ginger in a soup by cooking it too long, and you need to find a way to pull it forward at the end.

How do you know? *Taste.* Taste in the way that you've learned—to gather information and assess relationships. This assessing helps you understand what to do next, how to adjust. *And smell.* Your nose may pick up the excess of cumin wafting off a dish before your tongue does.

The seven dials can help you adjust. Remember that flavor is dynamic, and you might have to adjust more than once. Start with small amounts, adding a little bit at a time and tasting after each adjustment.

{ SALT }

I love you as meat loves salt," says the youngest daughter in an old English fairy tale that was the basis for *King Lear*. Her father doesn't properly appreciate the sentiment until he is served his meat without any. Lot knew better the power of salt—he turned his wife into a pillar of it, one of numerous appearances of salt in scripture. The substance was once so valuable that it was used as currency and, like some spices, incited wars. Since ancient times, it has been used to cure and preserve. The Egyptians used it as a means of keeping food and for embalming the bodies of the dead.

Poetic and mythic, ubiquitous and taken for granted, salt lives in contradictions. It's also, when you think about it, a strange and miraculous substance. There is salt that comes from the earth—it's the only rock we eat—and salt that comes from the sea. It is harvested in Hawaii and mined in the Himalayas, and farmed almost anywhere with a coastline, by dehydrating seawater.

Wherever it comes from, in whatever form, salt is the most important seasoning of all. Salt vivifies, draws out flavor, balances sweetness and acidity, increases aroma. Without the right amount of salt, the true flavor of a dish will not reveal itself.

Salt comes not only from different provenances but also in a range of sizes and shapes and distinct flavors. Salt can be more or less coarse, more or less minerally, flaky, or dense. Some salts are used for cooking and some for finishing. Because the intensity of salt varies considerably, for everyday salt find a brand that has a flavor you like and always use the same kind. This will allow you to be more consistent in your results.

At home three common varieties are useful. One is fine salt, either sea salt or kosher salt, for general seasoning. Fine salt is best for seasoning meat and fish, as it covers the surface evenly and dissolves in cooking. Flaky salt, like Maldon, is a nice finishing salt that is a little less intense. Added at the last minute, the delicate flakes dissolve on the tongue, asserting themselves and the flavors they boost more strongly than fine salt does. Coarse salts like sel gris or fleur de sel are more intense and rocklike, the thing to use when you want little bursts of saltiness. Fleur de sel comes in smaller grains and is good for all kinds of meats and vegetables. Sel gris is particularly good with sweeter vegetable dishes, like a salad of ripe tomatoes, when you want the salt to unfold as part of the dynamic experience of the dish.

While the "right" amount of salt is a somewhat subjective perception, generally speaking, you can judge this by the fullness of flavor of a dish overall. Pay attention as it evolves on your tongue, especially to how it ends. If it recedes too quickly, the dish probably needs more salt.

Salt does not work in a vacuum. It can be measured only in relation to the elements around it, in its specific context. Fatty meats need more salt; lean ones need less. Fat fixes flavors, but it also covers them, which means that fatty food generally requires more seasoning to balance flavors. Vegetables that are low in sodium or otherwise bland, like potatoes, need more salt, while those that are naturally high in sodium, like celery, need less.

Yet oversalting is far worse than undersalting. Too much salt overwhelms the flavors of the other ingredients, replacing them with the essentially one-dimensional taste of salt. The happy balance is the amount that draws out the flavors that are already present and fixes them in place.

Salt is most often used on its own, but it can also come from a salty cheese like Parmesan or a cured ingredient like anchovies or olives. Salt will not contribute any significant flavor of its own, but the other characteristics of the salty ingredient, of course, will.

Foods get sweeter as they cool and eventually freeze, but it's not because the sugars are going up. Sugar diminishes less quickly than salt, so perceptually colder food will seem sweeter, and hot food saltier.

Following are a few guiding rules on how salt interacts with other ingredients to fine-tune a dish.

Salt Rules

Salt pushes down sweet.

IF A DISH IS TOO SWEET, IT COULD BE THAT IT NEEDS A PINCH OF SALT. JUST A LITTLE BIT OF SALT HELPS MANY DESSERTS, NOT ONLY BY MAKING THEM LESS SWEET, BUT BY SHARPENING THE FLAVOR AS WELL. THINK ABOUT THE EFFECT THAT SALT HAS ON CARAMEL, PULLING OUT THE COMPLEX, BITTER COMPONENTS AND PUSHING DOWN THE SWEETNESS. OR IN A SIMPLE TOMATO SALAD, WHERE IT PUSHES AGAINST THE FRUITY SWEETNESS AND PULLS THE DISH INTO THE REALM OF VEGETABLES.

Salt pulls up sour.

ADDING SALT TO AN ACIDIC DISH MAKES THE ACIDITY MORE PROMINENT. SOMETIMES WHEN A SAUCE FEELS LIKE IT NEEDS MORE ACIDITY, WHAT IT REALLY NEEDS IS A PINCH MORE SALT. OR, IF A DISH IS HIGH IN ACIDITY, YOU MIGHT NOT WANT TO PUSH THE ENVELOPE ON THE SALT. THE RELATIONSHIP IS SYMBIOTIC.

Salt pushes down bitter.

SALT SPRINKLED ON RAW EGGPLANT DRAWS OUT BITTERNESS. THIS IS NOT A MATTER OF SEASONING, THOUGH; IT'S A MATTER OF CHEMISTRY. SALT DRAWS WATER OUT OF CELLS VIA OSMOSIS. IF A HIGHER CONCENTRATION OF SALT IS PLACED OUTSIDE THE CELL MEMBRANE, THE WATER WILL LEAVE THE CELL TO BOND WITH IT. THE LOSS OF WATER FROM THIS MOVEMENT WILL CAUSE THE PLANT CELLS TO SHRINK AND WILT.

SALT USED IN THIS WAY IS ACTUALLY A COOKING PROCESS——BY CREATING AN IMBALANCE BETWEEN THE SALT IN THE EGGPLANT'S CELLS AND THE EXTERIOR ENVIRONMENT, IT DRAWS MOISTURE OUT OF THE EGGPLANT AND ALONG WITH IT ANY RESIDUAL BITTERNESS. BUT EVEN WITHIN A DISH, SALT COUNTERACTS BITTERNESS: A SALAD OF RADICCHIO IS LESS BITTER WHEN SALTED, ESPECIALLY WHEN SALT IS ADDED IN CONJUNCTION WITH AN ACID, LIKE LEMON JUICE.

CONVERSELY, OVERSALTING CAN OFTEN BE MITIGATED BY ADDING ACID OR SWEETNESS.

Salted Cucumbers

Salting is a classic way of seasoning cucumbers in many countries, from Denmark to Japan. The salt draws out some of the cucumbers' moisture and, as with eggplants, also the bitterness, making them crunchier and bringing out brighter and fresher flavors. Salted cucumbers can be served as a side dish in their own right, stirred into a yogurt sauce, or seasoned with rice wine vinegar to make a quick pickle. The vinegar will diminish the salt, so they may require more seasoning.

SLICED OR DICED CUCUMBERS

SALT

Season the cucumbers with salt until they are pleasant to eat. Let them stand for at least 30 minutes.

Salted Caramel Sauce

Salted caramel is a great example of the alchemy of cooking. Just sugar, cream, and salt combine to create one of the most complex, beguiling dessert sauces you can make. The salt brings out the bitterness of the caramel, as well as the savory browned notes, and the cream smooths it all out and adds richness and luxurious mouthfeel. You can serve it drizzled over ice cream or cake, or cook chopped pears or apples in the finished sauce until tender, which will add gentle, fruity notes that round out the flavor.

2½ CUPS SUGAR

1 CUP WARM WATER

2 CUPS HEAVY CREAM

2 TEASPOONS SALT

Cook the sugar in a large pot until it caramelizes to a deep brown, 10 to 15 minutes. Add the water and stir until the sugar that clumps together dissolves. Add the cream and the salt and cook for 3 to 5 minutes, until the caramel is integrated with the cream. If not using immediately, store in the refrigerator for up to two weeks, reheating as needed.

Salt-and-Herb-Baked Celery Root

Cooking meat or fish encased in a crust of salt is a classic preparation. It also works with roots and vegetables, as in this recipe. The salt is mixed with herbs and moistened with egg white, which binds the salt into a hard crust that can be pulled off once the vegetable is cooked. Inside that crust the celery root will have steamed, to emerge moist, flavorful, and aromatic.

2 CUPS FINE SALT

½ TEASPOON FRESHLY GROUND BLACK PEPPER

2 EGG WHITES

½ CUP MINCED PARSLEY

2 TABLESPOONS MINCED FRESH THYME

2 TABLESPOONS MINCED CHIVES

2 TABLESPOONS MINCED FRESH TARRAGON

1 WHOLE CELERY ROOT, WASHED AND DRIED BUT NOT PEELED

Preheat the oven to 350 degrees.

In a food processor, blend the salt, black pepper, egg whites, and herbs until the mixture is green-colored. Put some of the mixture on a baking tray and put the celery root on top. Cover the celery root completely with the salt mixture. Bake for about an hour, until the celery root does not offer resistance when pierced with a cake tester or paring knife.

Remove from the oven and let stand for 10 minutes to cool a bit. Crack the crust and pull it away; discard. Peel the celery root, then slice and serve it. It will need a little salt.

{ SWEET }

Herodotus and Theophrastus and other ancient Greek writers had knowledge of raw sugar and called it "honey made by human hands, not that of bees." Bengal was probably the earliest region to manufacture sugar on any scale, giving it the Sanskrit name *sharkara*, which means "substance of small grains or stones." Sugar was integral to early pharmaceuticals. "Then as now sugar disguised the bitter taste of medicine but it also was useful as a way of preserving the often volatile ingredients of drugs," notes Paul Freedman in *Out of the East*. Sugar continued to be used in medicine, but this use also led to the development of confectionery. "Medicines were combined with sugar and by heating and cooling rendered into a variety of textures: gummy, hard, paste-like, soft or chewy," says Freedman. "These sugared medicinal preparations, known as 'electuaries,' are the origin of candy and many similar confections combining sugar and spices." The use of sugar helped popularize coffee, chocolate, and tea, making their natural bitterness palatable.

The sugar you are most likely to find in the store comes from sugarcane, a large grass that grows in subtropical regions. When the ripe cane is crushed and boiled, it produces a syrup that is dehydrated to make sugar.

What is interesting about sugar is that it has deep flavor in its "raw" state. Before processing, the sugar syrup is a rich brown, with the earthy-honeyed flavor that comes from molasses. Processing removes the molasses, which is sold separately, and creates several grades of sugar, from dark brown to light, Demerara (sticky golden crystals from the first crystallization stage of light cane juice) and turbinado (also a golden crystal from a raw sugar washed of its molasses) and the ubiquitous white sugar. White sugar is the rare ingredient that has no flavor to speak of, just the quality of sweetness.

Sweetness comes from many other sources, however—not only "sweeteners" like honey and agave and maple syrup but also fruits and vegetables like onions and beets and carrots. In each of those cases the sweetness is contained within a flavorful medium and lends the dish a fuller or leaner flavor, depending on the ingredients' degree of inherent sweetness.

In cooking, sweetness is a foil for salty, sour, and bitter, and useful for tempering them. But sweetness also carries a heaviness, so balancing foods with too much sugar can often make them dense and unpleasant. For example, duck glazed with honey is a classic combination, the inspiration for the Duck Breasts with Endive, Honey, Cinnamon, and Basil (page 167). But the proportion is important: too much of the honey will bury the bracing bitterness of the endive.

Sweetness can create perceptual richness, as in a vegetable stock or tomato soup. If the sugar is too low, the broth will seem watery and unsatisfying. The sweetness in the broth comes from the vegetables, like onions, carrots, and celery root. In a vinaigrette or other sauce, sweetness adds fullness. For example, honey in a vinaigrette balances the acidity in a heavier way than oil, creating a denser, more rounded dressing. In baking, sugar acts as a key chemical component, keeping breads and batters moist and helping to trigger the browning effect. But in sweet dishes it can also act much as salt does in savory food, as a seasoning that brings out flavor. For example, a pinch of sugar added to cut strawberries can make the strawberries taste stronger.

Sweet is special among the dials. It holds the center of many relationships—sweet and sour, sweet and salty, sweet and hot. Because of its heaviness and depth, it can temper and anchor sharp and angular tastes.

Sweet Rules

Sweet pushes down salt, sour, and bitter.

IF A DISH IS TOO SALTY, SOUR, OR BITTER, SWEETNESS CAN REDUCE THE EFFECT AND PROVIDE BALANCE.

SWEET IS ROUND AND SOFT, AND BLUNTS SHARPS EDGES. SWEET MAKES SALT AND ACID SOFTER AND LESS ANGULAR. ITS SOURCE COULD BE THE NATURAL SWEETNESS IN AN INGREDIENT LIKE ORANGE JUICE, BEETS, OR CORN, OR THE MORE DIRECT SWEETNESS OF HONEY OR SUGAR. THE SOURCE SHOULD MATCH THE DISH: THAT VINAIGRETTE WITH A TOUCH OF HONEY MIGHT BE TOO HEAVY FOR DELICATE YOUNG LETTUCE BUT JUST RIGHT FOR BALANCING HEARTY, BITTER RADICCHIO.

Sweet is heavy.

ADDING A PINCH OF SUGAR TO COOKED CARROTS CAN MAKE THEM TASTE MORE SUBSTANTIAL AND RICH. TOO MUCH, HOWEVER, CAN HAVE THE UNPLEASANT EFFECT OF WEIGHING DOWN THE FLAVOR. ADD SUGAR WHEN YOU WANT DEPTH.

Sweet acts like a seasoning in dessert.

WHEN A SWEET PREPARATION SEEMS THIN AND HOLLOW, IT MIGHT NEED MORE SUGAR. SUGAR BRINGS OUT THE FLAVOR OF SWEET DISHES, JUST AS SALT DOES WITH SAVORY FOOD.

Sweet-and-Sour Onion Condiment

This recipe uses the natural sweetness of onions, brought out by cooking, as the base for a sweet-sour condiment that makes a good accompaniment for pork or a tasty sandwich spread. The sherry vinegar has oxidized notes that lock with the lightly browned flavors of the onions. The condiment can accommodate a wide range of additional flavors, like raisins and capers as in this recipe, but also herbs, chili powder, or cumin and lime zest, folded in before refrigerating.

2 ONIONS, FINELY DICED

4 TABLESPOONS NEUTRAL-TASTING VEGETABLE OIL,
   SUCH AS GRAPESEED OR CANOLA

SALT

3 TABLESPOONS SHERRY VINEGAR

2 TABLESPOONS MINCED RAISINS

1 TABLESPOON MINCED CAPERS

FRESHLY GROUND BLACK PEPPER

Over medium-high heat, sauté the onions in the oil, stirring often, until tender and slightly browned. Salt them toward the end of this cooking process. Remove the pan from the heat and stir in the sherry vinegar, raisins, capers, and black pepper; adjust the salt. Cool, then transfer the jam to a jar or bowl. You may store it for up to two weeks in the refrigerator.

Sweet-and-Salty Braised Pork Belly

SERVES 6 AS A MAIN DISH, MORE AS PART OF AN ARRAY

It's hard to imagine another dish that packs in so much flavor with so little effort. Pork belly, the incredibly fatty cut that we know best in its cured state, as bacon (smoked) or pancetta (unsmoked), can take a lot of salt, spice, and umami. This recipe combines all those elements with complex fermentations in the form of soy and gochujang, a Korean condiment made of fermented chili, glutinous rice, soybeans, and salt. Together they lock into a flavor that's sweet, salty, aromatic, and deep.

The honey is a crucial ingredient. The sweetness buries the salt of the soy and provides depth, a base to accommodate the spicy fermented notes. The cooking method is a very slow, gentle overnight oven-steaming that allows the connective tissue in the pork to fully break down without interfering with the development of a complex aromatic bouquet.

3 POUNDS PORK BELLY

½ CUP SOY SAUCE

¾ CUP HONEY

2 TABLESPOONS *GOCHUJANG*

2 TABLESPOONS RED MISO

1-INCH KNOB GINGER, SLICED

1 TABLESPOON RICE WINE VINEGAR

Preheat the oven to 180 degrees. Place the pork in a baking pan. Whisk together the soy sauce, honey, *gochujang*, miso, and rice wine vinegar and pour the mixture over the pork, making sure to coat all sides. Scatter the ginger over everything and cover the pan with aluminum foil. Bake overnight, 12 to 15 hours, until the meat is fork-tender. When it is done, skim off the excess fat and reduce the remaining liquid until the salt level is correct. Then pour the liquid back over the pork.

Perfect accompaniments are sautéed greens and steamed rice as a neutral foil for the sauce, which needs no improvement.

Chilled Sweet and Hot Pepper Soup
SERVES 6

The pepper family is vast, but there are usually only a few kinds at the stores: bell peppers and chilies, red and green, sometimes yellow and orange Holland peppers as well. The difference between green bell peppers and their red, yellow, and orange counterparts is simply ripeness, and as with all fruits and vegetables, greater ripeness generally equates with greater sweetness.

This soup is a way of making the most of the ordinary peppers you will encounter in a grocery store. Here we want the sweetness of ripe red peppers, not the grassy, dusty taste of green ones. Although the ingredient list is short, the varied cooking methods bring out surprising depths of flavor. Some of the sweet peppers are charred, for a subtle smoky flavor. Some are boiled and blended and quickly chilled to preserve their sweetness. Some are juiced raw, which creates an incredibly sweet, red, fresh liquid. The sweetness of the soup is key to creating the depth that will balance the chili heat. A little bit of lime gives dimension and offsets the sweetness.

12 SWEET RED PEPPERS

3 HOT RED CHILIES

1 QUART VEGETABLE STOCK

FRESHLY SQUEEZED LIME JUICE

SALT

Char four of the red peppers over a flame until they are lightly blackened. Peel off their skins and remove the seeds and stems. Chop another four red peppers and the chilies and cook at a brisk boil in the vegetable stock for 10 to 15 minutes, until the peppers are softened. Blend all these peppers and pass them through a fine strainer into a bowl set into an ice bath.

Juice the remaining peppers in a vegetable juicer or blender and add them to the bowl. Strain the soup if any pieces remain. Taste. If the soup doesn't seem sweet enough, add a little sugar. Balance the flavor with lime juice and salt.

{ SOUR }

Sour describes anything with acidity. The acidity can come from many sources, from fermentations like vinegar to cultures like yogurt to acidic fruits, vegetables, and leaves.

Acidity balances all kinds of combinations. It cleans up murky, fatty flavors, brightens vegetables and soups, relieves richness, and generally energizes. Acidity is dynamic, and creates depth and complexity in otherwise flat flavors.

Vinegar is the strongest, most direct kind of sour, with piercing acidity. Red wine is the strongest vinegar, followed by sherry, white wine or champagne, and cider. Rice wine vinegar has the softest acidity, and balsamic has sweetness and a thick texture that balances the acidity and makes it more like a sauce.

Deciding which vinegar to use depends on the context. For a quick pickle of turnips, as in the recipe that follows, rice wine vinegar gives a gentle acidity that locks with the sweetness of the vegetable. The clean acidity of white wine vinegar might be just what's needed to cut through the dense taste of a rich sauce. Sherry vinegar adds an oxidized, nutty note, and balsamic has a fruity, sweet-sour flavor that connects well with fruit. Balsamic, which can also be used as a condiment for game or even on

Medieval Arab cuisine exemplifies the use of many of the dials as a complex way of orchestrating the final flavor and texture of a dish:

Many dishes derived their name and character from their acidic ingredients: sour milk or whey, vinegar, cider vinegar, lemon juice, bitter orange, pomegranate, crabapple, orange, apricot, or sumac berry. If the acidity seemed excessive, sugar could be added to correct it; honey was also used, but as often, sometimes with thick grape must, in order to obtain a sweet-and-sour taste. Dates, raisins, almonds, walnuts, and hazelnuts, shelled and chopped, not only sweetened the pot but thickened it as well.

—BERNARD ROSENBERGER, "Arab Cuisine and Its Contribution to European Culture"

a dessert, is a bit of a special case among vinegars. Because of its sweetness and balance, it is more of a flavoring than a source of acidity.

Citrus, from lemon to lime to grapefruit, has tempered acidity, with a mix of sweetness as well. Lemon is the most neutral and sweet, and it generally allows the focus to remain on the principal ingredients. Lime is angular and green, and locks with other green flavors, like cilantro or mint. Grapefruit is more bitter and dominant, with notes of pine and resin, and is a main player in any combination. Citrus changes in acidity quite a bit throughout the year, with the early picking in late fall being more acidic, becoming sweeter and riper as the season progresses. This progression is largely true for most fruit, from plums to tomatoes.

Some leafy vegetables and herbs are acidic to varying degrees, and some, like sorrel, have a high level of oxalic acid, and are both sour and green.

Fermented products—pickle-based condiments such as relishes, mustards, miso, hot sauce, and chili pastes, for example—can have some degree of acidity as well, and can be utilized if you are mindful of this quality. Cultured dairy products such as cheese, crème fraîche, and sour cream have a gentle, rich form of acidity from lactic acid. This acid comes enrobed in fat and is good in buttery sauces or salads where the goal is subtle adjustment.

Sour Rules

Sour energizes.

ACIDITY ADDS A DYNAMIC, ENERGETIC ELEMENT TO FLAT FLAVORS. A BLAND VINAI-GRETTE OR SAUCE CAN BE AWAKENED WITH A LITTLE VINEGAR OR LEMON JUICE. OFTEN, ACIDITY SEPARATES LAYERS OF FLAVOR, MAKING THEM MORE PERCEPTIBLE.

Sour ingredients can work together.

SOUR INGREDIENTS OFFER MORE THAN JUST SOURNESS. LEMON JUICE HAS SWEETNESS AS WELL, FOR EXAMPLE, AND HOT SAUCE HAS UMAMI. USING DIFFER-ENT KINDS OF ACIDITY IN THE SAME DISH IS A WAY TO CONTROL BALANCE AND FLAVOR. THE BLUEBERRY ICE WITH YOGURT AND LIME AND THE POTATO-SORREL SOUP, BELOW, EACH COMBINE PLANT ACIDITY WITH LACTIC ACIDITY. THE OLIVE-ANCHOVY VINAIGRETTE (PAGE 63) COMBINES THREE SOUR INGREDIENTS: VINE-GAR, LEMON JUICE, AND BRINED CAPERS.

Sour pulls down all the other dials.

ADDING ACIDITY DIMINISHES THE EFFECTS OF THE OTHER DIALS, IN DIFFERENT WAYS. SOUR COMBINED WITH SALT, SWEET, OR FAT REDUCES THEIR EFFECTS AND CHANGES THE BALANCE OF THE DISH. CUTTING AN UMAMI-RICH INGREDIENT LIKE MISO WITH VINEGAR NOT ONLY REDUCES THE SALTINESS OR THE MISO BUT DIMIN-ISHES THE DEPTH OF THE UMAMI, RENDERING THE CONDIMENT LIGHTER. ADDING SOUR TO HEAT, ON THE OTHER HAND, CAN SHARPEN AND FINE-TUNE THE HEAT, AS IN THE HOT SAUCE (PAGE 261).

Quick-Pickled Baby Turnips

This recipe can work with any thinly sliced vegetable that is good eaten raw, like carrots and radishes. With turnips it works especially well, as their slightly metallic edge is muted by the gentle acidity of the rice wine vinegar.

Here the vinegar pushes down the high amount of salt and sugar to create a fresh flavor. These are great as a garnish for a dish that needs a little bit of gentle acidity, like grilled pork chops and sautéed greens.

1 POUND BABY TURNIPS

1½ CUPS RICE WINE VINEGAR

1½ CUPS WATER

2 TABLESPOONS SUGAR

2 TEASPOONS SALT

Slice the turnips thinly and place them in a nonreactive bowl. In a pan, heat all the other ingredients to a simmer and pour the mixture over the turnips. Let cool to room temperature.

Potato-Sorrel Soup

Acidity can sometimes come from the central ingredients themselves, as in this potato soup enlivened with the green oxalic acidity of sorrel. Butter smooths it out, and a garnish of crème fraîche adds another layer of both fat and sour. Chives and black pepper provide punctuation. If you want less acidity, use half spinach and half sorrel.

1½ POUNDS RUSSET POTATOES

1 QUART VEGETABLE OR CHICKEN STOCK

4 TABLESPOONS BUTTER

SALT

1 BUNCH SORREL, COARSELY CHOPPED

4 TABLESPOONS CRÈME FRAÎCHE

2 TABLESPOONS CHOPPED CHIVES

FRESHLY GROUND BLACK PEPPER

Peel and thinly slice the potatoes. In a medium saucepan, bring the stock to a simmer and add the potatoes, butter, and a teaspoon or two of salt. Simmer until the potatoes are very tender.

Put the sorrel in the bottom of a blender pitcher, then add the cooked potatoes and liquid. Blend for a minute, until the mixture is thoroughly pureed and bright green. Cool quickly to preserve the pretty color: pour the hot soup into a metal bowl suspended in another bowl filled with ice and water, and stir constantly until cool. Adjust the salt.

To serve, heat the soup in a saucepan to just below a simmer. Divide among serving bowls. Drizzle the crème fraîche on top, and sprinkle with chives and black pepper.

Blueberry Ice with Yogurt and Lime

SERVES 4-6

Different kinds of acidity can lock, with the facets of one filling in the gaps in the other. One example is using both lemon and white wine vinegar in a vinaigrette. The lemon gives a soft, fruity acidity, and just a little vinegar can check the sweetness and create a more direct, agile acidity. If the proportion is inverted, the lemon can round out the aggressiveness of the vinegar.

Yogurt and lime have an affinity for each other. The green, floral notes of the lime zest lock with the creamy sourness of the yogurt, creating a merged, more complex flavor. The lime juice also lifts the blueberries, which can be sweet but flat, without changing their essential flavor. Here the acidity pushes down the sweetness and creates a more balanced dish.

½ CUP SUGAR

½ CUP WATER

5 CUPS FRESH BLUEBERRIES

2 TABLESPOONS FRESHLY SQUEEZED LIME JUICE

1 CUP WHOLE-MILK YOGURT

FRESHLY GRATED ZEST OF 1 LIME

SALT

Combine the sugar and water in a small saucepan and bring the mixture to a boil, stirring to dissolve the sugar. Let the mixture cool to room temperature. Place 4 cups of the blueberries in the pitcher of a blender or food processor and add the lime juice and the sugar syrup. The mixture will seem very sweet, but it will lose intensity as it freezes. Pour into a tray with sides at least 2 inches high and place in the freezer. Every 10 minutes, break up the ice crystals with a fork and return the tray to the freezer. When the mixture is fully frozen it should be light and a little creamy on the tongue, from the pectin in the blueberries.

When ready to serve, mix the yogurt with the lime zest. If the yogurt is not very sour, add a little lime juice as well. Add a pinch of salt to lift the acidity and aromatics.

For each serving, put a spoonful of yogurt in the bottom of a small bowl and top with a scoop of blueberry ice. Scatter with the remaining blueberries.

{ BITTER }

Unlike sweet, which we crave from birth, our taste buds must grow to enjoy bitter, perhaps in part because bitter flavors in the wild often indicate poison. In fact, many of the bitter flavors we most commonly eat are relatively recently domesticized wild plants. One example is radicchio. There are various kinds, each named after the town where they originated—Castelfranco and Treviso are but two. The cultivated versions were bred for enough sweetness to make them palatable, but they retain a bitter edge most palates mature to find pleasing.

Bitterness is most commonly found in vegetables—endive is another example— and less occasionally fruits and spices. Many bitter foods remain unpalatable for us, and for good reason: their bitterness is often a marker of badly grown or old produce, for example bolted lettuce or parched cucumbers.

But a slight, deliberately introduced bitterness is a welcome and even necessary element in flavor. We enjoy it in beer, tea, coffee, and grapefruit. Bitterness gives complexity and dimension to food that might otherwise be too plain. A little bitterness goes a long way toward balancing sweet flavors, or energizing bland ones.

So we think of bitterness, along with heat, as a special kind of dial—an accessory dial, to be deployed conservatively. The slight bitterness of light charring or grilling gives depth to and cuts the richness of fish, meat, and vegetables. A little bit of bitter greens can wake up an otherwise bland dish. (Conversely, an unintended bitter note, perhaps introduced by burning, can often be buried somewhat by a higher level of seasoning, for example by increasing the sweetness or acidity.)

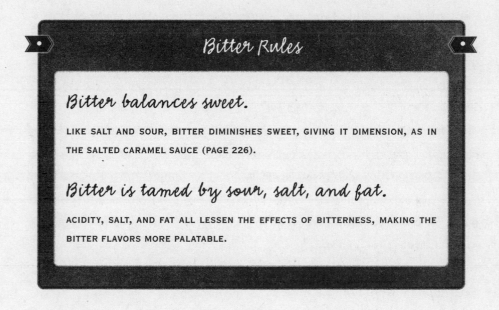

Bitter Rules

Bitter balances sweet.

LIKE SALT AND SOUR, BITTER DIMINISHES SWEET, GIVING IT DIMENSION, AS IN THE SALTED CARAMEL SAUCE (PAGE 226).

Bitter is tamed by sour, salt, and fat.

ACIDITY, SALT, AND FAT ALL LESSEN THE EFFECTS OF BITTERNESS, MAKING THE BITTER FLAVORS MORE PALATABLE.

Dandelion Greens with Radish and Blue Cheese Dressing

SERVES 4-6

This is a good salad to make converts out of people who are a little shy of bitterness. It is also a great illustration of how the dials work together to moderate one another.

Dandelion greens have a grassy, edgy bitterness that is diminished somewhat by cooking but needs the additional balance of acidity or fat. Here it gets both, in a dressing whose strong flavors also help bury the strong character of the greens. The dressing, made by blending blue cheese, mustard, oil, and vinegar, buries the greens in a few ways. The salt and the acid of the vinegar help pull down the bitterness. The mustard introduces its own note of acid and locks with the pungent blue cheese to make it more complex. Texture and proportion come into play too: The creaminess of the dressing coats the leaves thickly, giving the dressing the presence to hold its own against the assertive greens. The inclusion of endive, with its less intense bitterness and its watery crunch, tones down the intensity of the dandelion as well, and the spicy radishes lock with the black pepper to create a counterpoint.

2 OUNCES BLUE CHEESE

2 TABLESPOONS DIJON MUSTARD

6 TABLESPOONS CHAMPAGNE VINEGAR

8 TABLESPOONS NEUTRAL-TASTING VEGETABLE OIL,
SUCH AS GRAPESEED OR CANOLA

1 BUNCH YOUNG DANDELION GREENS

2 HEADS ENDIVE

1 BUNCH RADISHES, SLICED

SALT AND FRESHLY GROUND BLACK PEPPER

Blend the blue cheese, mustard, vinegar, and oil until smooth, adjusting the oil to control the intensity of the other ingredients. Put the dandelion greens in a salad bowl and tear them into bite-size pieces. Separate the endive into leaves, tearing them if they are large, and add them too. Add the radishes. Add the dressing and toss until the leaves and radishes are lightly coated. Season with salt and black pepper. If you have extra dressing, it's great with grilled chicken, as a complement to the salty, smoky meat.

Charred Eggplant Relish

This relish is a simpler cousin of baba ganoush. You can add whatever spices you like, but in this stripped-down form the focus is on the flavor of the charred eggplant. A slightly bitter vegetable gets an overlay of bitter, almost burnt flavors from being charred in a pan or over a grill. But when the skin is removed, it reveals sweet, smoky, flavorful flesh, and the gentle acidity of the rice wine vinegar pulls back the bitter notes. Cheerful mint locks with oniony scallion, and the different facets of the herbs melt into one bright, green lifting note. The black pepper has an important role too, punctuating the middle notes and locking with the earthy charred flavors.

Serve as a dip or a spread for flatbread, or to accompany grilled chicken or fish.

4 LARGE JAPANESE EGGPLANTS

1 TABLESPOON VEGETABLE OIL

2 TABLESPOONS RICE WINE VINEGAR

1 TABLESPOON CHOPPED MINT

1 TABLESPOON THINLY SLICED SCALLION, GREEN AND WHITE PARTS

1 TABLESPOON FRUITY OLIVE OIL

SALT AND FRESHLY GROUND BLACK PEPPER

Coat the outside of the eggplants with the vegetable oil and char over a grill or in a hot cast-iron pan until the outside is blackened and the inside is soft. Cool the eggplants, then cut them in half and scoop out the flesh. Roughly chop and combine with the vinegar, mint, scallion, and oil. Season with salt and black pepper.

Bitter Chocolate – Espresso Mousse

SERVES 4

Here the fat in the eggs and the cream balances the bitterness of the chocolate and coffee, burying their sharp edges and making the mousse round and harmonious.

8 OUNCES 70 PERCENT CACAO CHOCOLATE

½ CUP STRONG COFFEE

4 EGGS

4 TABLESPOONS SUGAR

1½ CUPS CHILLED HEAVY CREAM

In a heatproof mixing bowl placed over (not in) a pot of simmering water, melt the chocolate with the coffee, stirring frequently. Remove the bowl from the heat.

Combine the eggs and sugar in a separate bowl and put it over the simmering water. Whisk constantly until the mixture is light yellow and forms thick ribbons when you lift the whisk. Remove the bowl from the heat, add the melted chocolate, and stir to combine. Cool the mixture to room temperature.

In a clean mixing bowl, whisk the cream to soft peaks. Gently fold half into the chocolate mixture until it is thoroughly blended. Fold in the rest, then transfer the mousse to individual serving bowls (or one large bowl) and chill it in the refrigerator for at least 4 hours. Serve cold.

{ UMAMI }

S avoriness, or the glutamic taste, has been consciously used for hundreds of years in Japan. According to Ole G. Mouristen and Klavs Styrbæk, authors of *Umami: Unlocking the Secrets of the Fifth Taste*, it was Kikunae Ikeda, a Japanese chemist, who in 1908 discovered monosodium glutamate (MSG) as the source of the distinctively delicious taste of the widely used Japanese soup stock dashi, which is based on shaved bonito flakes. "He was the first person to investigate this taste in a scientific manner and introduced the term by which it is now known: umami," they write. Seaweed has the taste of umami, as do tomatoes and mushrooms. Broccoli, Parmesan, aged cheese, blue cheese, walnuts, fish, and shellfish are ingredients high in the taste of umami. Fermented foods like sourdough bread, miso, and soy sauce also contain high levels.

Yet umami is not a flavor per se but, as with the other dials, a quality—in this case, an intensity of flavor, an amplification or a concentration, something akin to the effect that salt has on food. Add ingredients high in umami to create a dish with depth and power. Umami can also lift a dish that otherwise feels too light or insubstantial. When you add condiments like ketchup, Worcestershire sauce, or pickles to your burger, you're reaching for umami—for boldness.

Umami can be introduced through ingredients that have already been processed or engineered to create it (aged meats and cheeses, for example) or created through cooking techniques (the browning that comes with sautéing, searing, and roasting). Soups and stocks allow the locking of ingredients that can create umami, as in a classic dashi, which unites *katsuobushi* (dried, smoked, fermented tuna) and seaweed. Umami can be heightened with the addition of salt or diminished through acidity. Too much umami is not necessarily a good thing. It can create a flavor that is overly dense and overwhelming. Combining flavors with unequal levels of umami can lead to unwanted burying. For example, the sweetness and delicacy of boiled carrots might be buried when combined with heavily roasted meats.

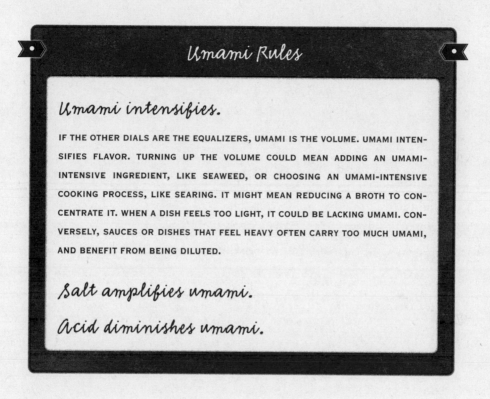

Umami Rules

Umami intensifies.

IF THE OTHER DIALS ARE THE EQUALIZERS, UMAMI IS THE VOLUME. UMAMI INTEN-SIFIES FLAVOR. TURNING UP THE VOLUME COULD MEAN ADDING AN UMAMI-INTENSIVE INGREDIENT, LIKE SEAWEED, OR CHOOSING AN UMAMI-INTENSIVE COOKING PROCESS, LIKE SEARING. IT MIGHT MEAN REDUCING A BROTH TO CON-CENTRATE IT. WHEN A DISH FEELS TOO LIGHT, IT COULD BE LACKING UMAMI. CON-VERSELY, SAUCES OR DISHES THAT FEEL HEAVY OFTEN CARRY TOO MUCH UMAMI, AND BENEFIT FROM BEING DILUTED.

Salt amplifies umami.

Acid diminishes umami.

Tofu Braised with Seaweed

SERVES 4

Foods rich in umami work well to flavor broths and bland foods. Here the seaweed and mushrooms lock into a rich, deep base for a broth.

½ POUND FRESH SHIITAKE MUSHROOMS, COARSELY CHOPPED,
  PLUS 1 CUP SLICED FOR GARNISH

5 CUPS WATER

3-INCH STRIP DRIED KONBU SEAWEED

SALT

1 CUP SLICED LEEK

1 CUP SLICED CARROT

1 CUP SLICED CABBAGE

12 OUNCES FIRM TOFU, CUT INTO ½-INCH CUBES

½ TEASPOON *SHICHIMI TOGARASHI* (A JAPANESE SPICE BLEND—CHILI FLAKES
  ARE A GOOD SUBSTITUTE)

2 TABLESPOONS GRATED GINGER

2 TABLESPOONS SOY SAUCE

Simmer the chopped shiitakes in the water for 1 hour, covered, then add the konbu and simmer for 30 minutes more. Strain the broth and return it to the pot, reserving the cooked konbu; discard the shiitakes. Bring the broth to a simmer again and adjust the salt.

Julienne the cooked konbu and add it to the pot, along with all the remaining ingredients except the ginger, soy sauce, and sliced shiitakes.

When the vegetables are tender, remove the pot from the heat and add the ginger and soy sauce. Add the sliced shiitakes and let stand for 5 minutes. Adjust the seasonings and serve.

Poached Celery with Mustard Dressing and Parmesan

SERVES 4-6

This is basically a version of Celery Victor, the classic San Francisco salad of poached celery with mustard, just amped up with umami. The dressing has a bit of anchovy, and the finished dish is showered in grated Parmesan. If you have crisp, flavorful celery from the farmers' market, consider making this salad with shaved raw celery. But if your vegetable has been languishing in the back of the refrigerator, proceed as directed; the added umami will make it delicious.

1 HEAD CELERY

1 TABLESPOON DIJON MUSTARD

2 TABLESPOONS WHITE WINE VINEGAR

3–4 TABLESPOONS NEUTRAL-TASTING VEGETABLE OIL, SUCH AS GRAPESEED OR CANOLA

6 ANCHOVY FILLETS, MINCED

SALT AND FRESHLY GROUND BLACK PEPPER

1 CUP FRESHLY GRATED PARMESAN

Separate the celery into stalks, and wash, trim, and peel them. Simmer them in salted water until tender. Drain, then chill the celery in the refrigerator. While the celery is cooling, whisk together the mustard, vinegar, oil, and anchovies, and season to taste with salt and black pepper.

To serve, line up the cooked celery, drizzle it generously with the mustard dressing, and top it with the Parmesan.

Charred Broccoli with Misonnaise and Black Sesame

SERVES 4

Broccoli is often buried under a fatty sauce, like hollandaise or mornay, perhaps because some people don't like its flavor very much. Broccoli is delicious with creamy sauces, but its deep green, flat flavor could use a lift from some acidity. It also benefits from a strong counterpoint, like olives or miso. Here red miso, with its earthy, dense, salty notes, adds umami and depth to cooked broccoli.

Because miso is so strong, the broccoli requires a preparation that intensifies and concentrates its flavor to match the intensity of the miso. Cooking the broccoli in salted water makes it seasoned, sweet, and tender; charring the outside in very hot oil gives it a smoky, bitter edge and creates complex secondary aromatics and umami notes that complement the fermented flavors of the miso.

The strength of the miso needs to be diluted in some way. Since broccoli goes so well with creamy sauces, what about blending the miso into a mayonnaise—a misonnaise?

This dish has a lot going on flavor-wise, and doesn't need much to complete it. A bit of rice wine vinegar provides the needed acidity. Here it's added to the broccoli, not the misonnaise, so it doesn't interfere with the miso's umami qualities. The broccoli is dressed with the vinegar just after it has cooked in water, when the vinegar is easily absorbed—its brightness in effect cooked into the vegetable rather than outside it.

The combination could benefit from one finishing note. The flavors are so rich and complex that an herb might distract, but the bitter/earthy flavor of toasted black sesame offers a textural and flavor counterpoint to both the vegetable and the sauce.

¼ CUP RED MISO

1 EGG

3 TABLESPOONS WATER

1¼ CUPS VEGETABLE OIL

1 TABLESPOON BLACK SESAME SEEDS

1 HEAD BROCCOLI

SALT

2 TABLESPOONS RICE WINE VINEGAR

Put the miso, egg, and water in a blender. Blend at medium speed, drizzle in 1 cup of the vegetable oil, and blend until the mixture is thick and creamy. Taste for seasoning—likely the miso will provide sufficient salt. Refrigerate.

Toast the black sesame seeds in a 250-degree oven, stirring or shaking often, for about 20 minutes. (A toaster oven is good for this.) Be careful not to burn the seeds—you're waiting for the smoky, dark qualities of the sesame to emerge, but you don't want them to become bitter.

Peel the broccoli stems and cut the crowns and stems into large spears. Bring a large pot of water to a boil, salt well, and add the broccoli. Cook until tender, drain well, and then toss with the rice wine vinegar. Cool on a tray.

Heat a cast-iron pan or griddle to smoking. Toss the broccoli with the remaining vegetable oil and some salt, and transfer to the pan or griddle. Cook, turning often, until irregularly blackened.

Divide the broccoli among serving plates. Drizzle generously with the misonnaise, and sprinkle with the toasted sesame seeds.

{ FAT }

Recent research by Richard Mattes, a professor of foods and nutrition at Purdue University, indicates that humans can indeed taste fat. As reported in the July 2015 issue of the journal *Chemical Senses*, his team demonstrated a specific "oleogustus," as distinct from other basic taste responses (sweet, sour, salty, and bitter).

The function of fat within a dish is to carry and disburse flavor. It modifies texture and mouthfeel, and therefore the release of flavor. It also balances acidity, as in a vinaigrette. It can be symbiotic with umami, as in cheese. Generally speaking, fat is a blanket that lies on top of highly flavored ingredients, keeping them down. It has this

The pastoral peasant tradition of the Near East contributed the oil in which almost every Baghdad dish was put to cook—*alya*, the fat rendered from sheep's tails. Time after time al-Baghdadi [author of a famous ancient cookbook] began his instructions with the words, "Cut meat into middling pieces; dissolve tail and throw away the sediment. Put the meat into this oil and let it fry lightly." The popularity of tail fat may have had something to do with the existence of the local fat-tailed sheep, though whether as cause or effect remains a matter for debate.

—REAY TANNAHILL, *Food in History*

effect on the other dials as well; adding fat often means ratcheting up the other dials, especially salt and acid.

Fat comes in many forms—animal fat, present in meat, poultry, and fish, and also on occasion used independently of their flesh (duck fat, chicken fat, suet, lard). It's also present in animal products—eggs, milk, cream, butter. And it is found in oils pressed from olives, legumes, and nuts.

Fats are all distinctive ingredients with their own complex flavors. Animal fats are heavy and thick and can be used to enrich not only meat dishes but vegetable dishes as well, acting as both a medium for heat and a flavor booster, as in potatoes fried in duck fat. Dairy fats, especially butter, are an incredibly versatile cooking medium that impart sweetness as well as richness. For a leaner profile, as in a salad dressing, vegetable and nut oils work best.

Fat Rules

Fat fixes flavor.

WHEN FLAVORS ARE INTRODUCED INTO FAT, THROUGH BLENDING, COOKING, OR INFUSING, IT CAN BE A VEHICLE FOR REINFORCING FLAVOR. FAT ABSORBS AND HOLDS ALL THE FLAVOR ELEMENTS AROUND IT, AND CAN AMPLIFY FLAVOR, AS IN CRAB BUTTER (PAGE 254) OR SMOKED OIL (PAGE 197).

Fat pushes down sour, salt, bitter, and heat.

HOWEVER, FAT THAT IS INTRODUCED AT THE END OF COOKING, WHEN THE FLAVOR LOCKS HAVE BEEN ACHIEVED, CAN HAVE THE OPPOSITE EFFECT. FAT USED THIS WAY CAN LIE ON TOP OF OTHER FLAVORS AND SUPPRESS THEM. FAT CAN BALANCE ACIDITY AND CAN PUSH DOWN SALT AND BITTER. ADDING FAT CAN MAKE OVERLY SALTY OR BITTER FLAVORS LESS PROMINENT. FAT ALSO BRINGS DOWN THE PERCEPTION OF SPICY HEAT.

Crab Butter

This recipe recalls one of the most time-honored ways to eat crab—with drawn butter. The sweet dairy fat wraps around the briny, umami notes of the crab, and the crab flavor becomes fixed in the fat. In the same way, simmering shells in butter as it clarifies creates an infused crab butter, a perfect marriage. This butter can be drizzled over a crab sauce or bisque to amplify its flavor, and it makes a delicious sauce for a piece of steamed fish.

2 TABLESPOONS VEGETABLE OIL

SHELLS FROM 1 COOKED CRAB, SMASHED, NO INNARDS

1 POUND BUTTER

Heat the vegetable oil over high heat in a saucepan, and add the shells. Cook for 3 to 4 minutes, stirring often, until aromatic and lightly browned. Reduce the heat to low and add the butter. Cook at a low simmer until the butter turns clear. Strain the butter and store it in the refrigerator or freezer; it will keep one week in the refrigerator, and if well wrapped, months in the freezer.

Poached Eggs with Greens and Potatoes Cooked in Bacon and Butter

SERVES 4-6

Combining fats can have multiple effects. One is that a more neutral-tasting fat, like butter, becomes a vehicle to stretch a more intensely flavored fat, like smoked pork fat (in the form of bacon)—a way to increase the overall level of fat without making the flavor overwhelming. The bacon cooks in the butter, thinly sliced potatoes are added and cook until halfway done, and then hearty greens like kale are added. The pot is covered and the greens cook in the fat and in only as much water as is bound in their cells and clings to them from being washed. Sunchokes, if you can find them, add a nutty sweetness that connects with the earthy facets of the potato and balances the dark bitter notes of the kale. The whole takes on a deep, persistent flavor very different from the flavor of greens simply simmered in water. This is a great approach to greens that are a little tired and could use a flavor boost. The yolks from the poached eggs on top, running into the greens, add one more fat element.

2 TABLESPOONS BUTTER

2–3 STRIPS BACON, CUT INTO ½-INCH SLICES

1 POUND YUKON GOLD POTATOES, WASHED AND
SLICED ¼ INCH THICK

SALT

1 BUNCH KALE, WASHED AND SLICED INTO 1-INCH STRIPS

½ POUND SUNCHOKES, PEELED AND THINLY SLICED

FRESHLY GROUND BLACK PEPPER

2 TABLESPOONS WHITE WINE VINEGAR

8 EGGS

In a large, heavy saucepan, melt the butter over medium heat. Add the bacon and cook until almost cooked through but not crisp. Add the potatoes and a little salt and continue to cook, turning often, until the potatoes are lightly browned and halfway tender, about 8 to 10 minutes. Add the kale and sunchokes and a bit more salt, cover, and cook over medium-low heat, stirring often and adding a bit more water if necessary, until the kale is tender, about 10 to 15 minutes. Season with black pepper and adjust salt.

While the potato mixture is cooking, fill a wide, deep sauté pan with water to a depth of at least 3 inches and bring to a rolling boil. Salt well. Drop in the eggs as quickly as you can, spacing them apart. Turn off the heat, cover, and let stand for exactly 5 minutes.

Divide the potato mixture among individual bowls, top with a poached egg or two, and grind a little more pepper over the top.

Chopped Salad with Avocado and Spicy Vinaigrette

SERVES 6-8

This bright, summery salad is a good way to see how fat balances acid and heat. You can make it with whatever vegetables you have on hand, raw or cooked, as long as you have some hearty greens like romaine to provide a crisp, clean base. The vinaigrette combines two kinds of fat: flavorless vegetable oil, whose effect is only to connect and temper the flavorful ingredients, and a bit of fresh cheese to smooth and enrich the dressing, allowing it to stand up to the vegetables.

1 RIPE TOMATO, SKINNED AND ROUGHLY CHOPPED

2 TABLESPOONS CHAMPAGNE VINEGAR

2 TABLESPOONS FRESHLY SQUEEZED
   LIME JUICE

1 SERRANO CHILI, SEEDED

2 TABLESPOONS NEUTRAL-TASTING VEGETABLE OIL,
   SUCH AS GRAPESEED OR CANOLA

1 TABLESPOON FRESH GOAT'S OR COW'S MILK CHEESE

SALT

2 HEADS ROMAINE LETTUCE, CUT INTO 1-INCH PIECES

2 CUPS SUMMER VEGETABLES (SQUASH, BELL PEPPERS, CUCUMBERS,
   FENNEL, ETC.), CHOPPED, SHAVED, OR COOKED AND CUT
   INTO BITE-SIZE PIECES

1 AVOCADO, DICED

FRESHLY GROUND BLACK PEPPER

Place the tomato, vinegar, lime juice, chili, oil, and cheese in the pitcher of a blender with a little salt and process until smooth.

Combine the remaining ingredients in a serving bowl. Dress with the tomato vinaigrette and season to taste with salt and black pepper.

{ HEAT }

We consider heat the seventh dial because of the way that spiciness is perceived in the mouth. Heat creates dynamic, complex flavor. It can invigorate bland foods, alleviate richness, and bring excitement to otherwise homogenous combinations. It intensifies sharp spices, as if they themselves are generating the heat. Adding heat along with other flavor ingredients—such as cayenne along with cardamom or cinnamon—can create a lock that renders the cardamom or cinnamon more intense.

Heat Rules

Heat creates dynamic, complex flavors.

Heat alleviates richness.

HEAT IS AN IMPORTANT ELEMENT IN ALMOST EVERY CUISINE, FROM THE GENTLE, DUSTY SPICE OF WHITE PEPPER TO THE PUNGENCY OF MUSTARD TO THE FIRE OF A HABANERO. ADDING HEAT MAKES FLAVORS MORE VIVID AND ENERGETIC, IN PART BECAUSE OF THE PHYSICAL REACTION OUR PALATES HAVE TO HEAT. THIS ENERGY HELPS HEAT CUT THROUGH RICH, FATTY FLAVORS.

Heat intensifies spices.

INGREDIENTS THAT CONTRIBUTE HEAT LOCK WITH SPICES, AMPLIFYING THEIR EFFECTS. A CHILI IN A MILDLY SPICED DISH WILL BRING THE SPICES TO THE SURFACE.

Heat works well with every other dial.

LIKE SWEETNESS, HEAT IS A DANCES-WITH-EVERYONE PARTNER THAT DOESN'T STEAL CENTER STAGE. ADD A SMALL AMOUNT OF MUSTARD TO A VINAIGRETTE AND IT'S STILL A VINAIGRETTE, JUST A TOUCH MORE COMPLEX. A LITTLE BIT OF CHILI CAN LOCK WITH THE UMAMI OF A PORK ROAST.

Heat comes not just from spices (the various forms of pepper and mustard, for example), but also from fruits like chilies (yes, a fruit!) and from plants like ginger and horseradish. The heat can be subtle or intense, apparent immediately or released slowly. If it's hard to imagine how to use spiciness across a spectrum, consider the most common spicy flavor—black pepper—and the different sensations it creates when used sparingly or heavily, finely or coarsely ground.

A cuisine intimately knowledgeable about all varieties of chili peppers is Mexico's. In taking a closer look at actual culinary practice in a Mexican village, Elizabeth and Paul Rozin detailed the many ways that chili pepper is used:

What seems a monotonous repetition of flavorings may well be an illusion of the outsider, just as all red burgundies taste alike to the inexperienced wine drinker. A closer look at actual culinary practice reveals a rich and subtle variation of flavoring from dish to dish and from meal to meal. In the Mexican village we studied, for example, chili pepper is used in many ways. It is cooked into soups and stews or sliced and placed upon other foods or ground into a sauce, often with tomatoes and other ingredients. At least ten types of chili peppers are used in this village: some are fresh, some are dried, some are red, others green, some strongly piquant, others less so. They differ somewhat in the type of "burn" they produce in the mouth. And villagers assure us that each type of chili has a different taste. . . . Within the single "theme" of this Mexican chili pepper flavor principle there is a great deal of subtle and controlled variation.

—ELIZABETH AND PAUL ROZIN, "Culinary Themes and Variations"

Hot Sauce

Chili heat is sharp, but not very complex. Adding acidity gives life to a hot sauce, and thickening it gives more presence in the mouth and allows it to emulsify the ingredients in the sauce. This is a great way to preserve chilies at the height of ripeness, and it will create a more distinctive and flavorful sauce than you can buy commercially. And it's fun to make!

1 POUND FRESH RED CHILIES, LIKE RED FRESNO OR JALAPEÑO

2 CUPS CHAMPAGNE VINEGAR

2 TABLESPOONS SALT

1 TEASPOON XANTHAN GUM

Halve the chilies lengthwise and remove and discard the seeds. Combine the pods with the vinegar and salt in a nonreactive container (a glass jar is best). Cover and let them stand in a refrigerator for at least a week, then transfer them to a blender or food processor. Add the xanthan gum, and process on high until the mixture is smooth. Return it to the jar and store it in the refrigerator. It will last a few weeks.

Spiced Red Lentil Stew
SERVES 4

In this stew of earthy lentils, the cayenne locks with the other spices to amplify their effect. This layering happens by binding them into the onion, garlic, and oil at the beginning, and then the new merged taste flavors the soup. This is great with a dollop of yogurt (acidity and dairy to temper the heat) and a little lime zest (a green note to brighten it and layer the finishing aromatics). Serve over rice.

1 MEDIUM ONION, DICED

3 TABLESPOONS VEGETABLE OIL

SALT

2–3 CLOVES GARLIC, MINCED

1 TEASPOON GROUND CINNAMON

5 PODS CARDAMOM

1 TEASPOON DRIED GROUND GINGER

2 TEASPOONS GROUND CUMIN

½ TEASPOON CAYENNE

2 CUPS RED LENTILS

6 CUPS WATER

In a soup pot, cook the onion in the oil over low heat with a little salt until it is tender. Add the garlic and cook for 2 minutes more. Then add the spices and cook, stirring, for a few minutes more, until they are aromatic but not in danger of burning. Add the lentils, water, and a couple of teaspoons of salt. Bring to a boil, cover, and simmer until the lentils are tender, about 20 minutes.

Orange with Chili Salt

This combination is a street food popular in Mexico. The sweet-sour orange is balanced with salt and lifted by the chili, whose fruitiness also locks with the floral facets in the orange.

NAVEL ORANGES

POWDERED CHILI, SUCH AS PIMENT D'ESPELETTE, ALEPPO, OR CAYENNE

SALT

Peel the oranges and slice them thickly. Mix together the chili and salt, sprinkle over the orange slices, and serve.

ACKNOWLEDGMENTS

I want to thank my coauthor and dear friend Daniel Patterson for the sheer joy of working together to find words for the wordless process of creating flavor. I am grateful for his brilliance, kindness, originality, and creativity, and for our loving partnership. I am blessed to call Harold McGee a close friend and am so grateful for his support, kindness, and invaluable feedback on our manuscript. I have such deep love and gratitude for my precious son, Devon Curry, who scanned my oodles of research and gracefully helped with absolutely Everything. I love his daily presence in my life here at Aftelier.

I want to thank my best friend and editor, Becky Saletan, for all the work she put into every page—every sentence—of this book, and for more than twenty years of deep and loving intimacy. And I thank my husband, Foster Curry, for his endless note-taking and pitching in in every possible way to bring this book into the world. Foster, the love of my life, thank you for making each day of my life so happy and filled with joy.

—M.A.

I would like to join Mandy in thanking Harold McGee for his insights and feedback, and for creating a body of work that has informed so many of our ideas.

Thanks as well to Rebecca Saletan for your brilliant and hard work, keen insight, and unerring instincts in helping to turn a promising idea into a finished book. I was blown away, over and over, by everything you contributed.

Thank you to Foster Curry, for your incredible ideas, your hard work, your note-taking, your editing, and for always bringing positive energy to every conversation. It was wonderful to have you as part of the process.

And most of all, thank you to Mandy, without whom this book never would have been started, and certainly never would have been finished. You did so much, from research to developing concepts to structure to making sure I stayed on track (not an easy task!). I see your influence, hard work, and caring in every sentence. Thank you for teaching me so much. I have no words to express how deeply indebted I am to you.

—D.P.

Thanks also from us both to the amazingly creative, dedicated, and meticulous people at Riverhead: publisher Geoff Kloske for his support and attention; the design team, Helen Yentus, Claire Vaccaro, Amanda Dewey, and Grace Han, for the gorgeous package inside and out; Maureen Klier and Anna Jardine for deft copyediting and tweaking; Katie Freeman, Glory Plata, Jynne Dilling, Kate Stark, and the rest of the publicity and marketing crew for working tirelessly and imaginatively to bring before the world a book that is incredibly dear to our hearts; and Wendy Pearl and the rest of the Penguin sales team for spreading the gospel.

BIBLIOGRAPHY

Albala, Ken. *Eating Right in the Renaissance*. Berkeley: University of California Press, 2002.

Ashurst, Philip R., ed. *Food Flavorings*. 2nd ed. Dordrecht: Springer Science+Business Media, 1991.

Bourdieu, Pierre. "Taste of Luxury, Taste of Necessity." In Korsmeyer, *The Taste Culture Reader*, 72–78.

Brillat-Savarin, Jean Anthelme. *The Physiology of Taste: Meditations on Transcendental Gastronomy*. Translated and edited by M. F. K. Fisher. New York: Vintage Books, 1949.

Chartier, François. *Taste Buds and Molecules*. Translated by Levi Reiss. Hoboken, NJ: John Wiley & Sons, 2012.

Civitello, Linda. *Cuisine and Culture: A History of Food and People*. Hoboken, NJ: John Wiley & Sons, 2011.

Classen, Constance. *Worlds of Sense: Exploring the Senses in History and Across Cultures*. London: Routledge, 1993.

Classen, Constance, David Howes, and Anthony Synnott. "Artificial Flavours." In Korsmeyer, *The Taste Culture Reader*, 337–42.

Dalby, Andrew. *Tastes of Byzantium: The Cuisine of a Legendary Empire*. London: I. B. Tauris, 2010.

Daumal, René. *Rasa, or, Knowledge of the Self*. New York: New Directions, 1982.

David, Elizabeth. *Spices, Salt and Aromatics in the English Kitchen*. London: Grub Street, 2000.

De Pascalis, Andrea. *Alchemy: The Golden Art*. Rome: Gremese International, 1995.

Dorland, Wayne E., and James A. Rogers, Jr. *The Fragrance and Flavor Industry*. Mendham, NJ: Wayne E. Dorland, 1977.

Eamon, William. *Science and the Secrets of Nature: Books of Secrets in Medieval and Early Modern Culture.* Princeton, NJ: Princeton University Press, 1994.

Etkin, Nina. *Edible Medicines: An Ethnopharmacology of Food.* Tucson: University of Arizona Press, 2006.

Flandrin, Jean-Louis. "Seasoning, Cooking, and Dietetics in the Late Middle Ages." In Flandrin and Montanari, *Food: A Culinary History*, 313–27.

———. "From Dietetics to Gastronomy: The Liberation of the Gourmet." In Flandrin and Montanari, *Food: A Culinary History*, 418–32.

Flandrin, Jean-Louis, and Massimo Montanari, eds. *Food: A Culinary History.* Translated by Albert Sonnenfeld. New York: Columbia University Press, 1999.

Flückiger, Friedrich A., and Daniel Hanbury. *Pharmacographia: A History of the Principal Drugs of Vegetable Origin, Met with in Great Britain and India.* London: Macmillan, 1879.

Freedman, Paul. *Out of the East: Spices and the Medieval Imagination.* New Haven, CT: Yale University Press, 2008.

———, ed. *Food: The History of Taste.* Berkeley: University of California Press, 2007.

Gilchrist, Cherry. *The Elements of Alchemy.* London: Element, 1991.

Goswamy, B. N. "*Rasa*: Delight of the Reason." In Korsmeyer, *The Taste Culture Reader*, 215–25.

Harrop, Joseph. *Monograph on Flavoring Extracts: With Essences, Syrups and Colorings.* Columbus, OH: Harrop, 1891.

Hudson, W. H. *A Hind in Richmond Park.* London: J. M. Dent & Sons, 1922.

Ingold, Tim. *Making: Anthropology, Archaeology, Art and Architecture.* New York: Routledge, 2013.

Jackson, Ronald S. *Wine Tasting: A Professional Handbook.* London: Elsevier, 2009.

Johnston, James F. *The Chemistry of Common Life.* New York: D. Appleton, 1855.

Jullien, François. *In Praise of Blandness.* New York: Urzone, 2004.

Kaiser, Roman. *Meaningful Scents Around the World.* Zurich: Wiley-VCH, 2006.

Korsmeyer, Carolyn, ed. *Making Sense of Taste.* Ithaca, NY: Cornell University Press, 1999.

———. *The Taste Culture Reader: Experiencing Food and Drink.* New York: Berg, 2005.

La Varenne, François Pierre. *La Varenne's Cookery: The French Cook; The French Pastry Chef; The French Confectioner.* Translated and edited by Terence Scully. Blackawton, England: Prospect Books, 2006.

BIBLIOGRAPHY

Laudan, Rachel. *Cuisine and Empire: Cooking in World History*. Berkeley: University of California Press, 2013.

Leyel, Mrs. C. F. *The Magic of Herbs: A Modern Book of Secrets*. London: Butler and Tanner, 1926.

Maarse, Henk, ed. *Volatile Compounds in Foods and Beverages*. New York: Marcel Dekker, 1991.

Matthews, A. C. "Beverage Flavourings and Their Applications." In Ashurst, *Food Flavorings*, 160–86.

May, Robert. *The Accomplisht Cook, or the Art and Mystery of Cookery*. Gloucester, England: Dodo Press, 2010.

McFadden, Christine. *Pepper: The Spice That Changed the World*. Bath, England: Absolute Press, 2008.

McGee, Harold. *On Food and Cooking*. Rev. ed. New York: Scribner, 2004.

Merory, Joseph. *Food Flavorings: Composition, Manufacture, and Use*. Westport, CT: AVI, 1960.

Miller, H. D. "The Pleasures of Consumption: The Birth of Medieval Islamic Cuisine." In Freedman, *Food: The History of Taste*, 135–61.

Mintz, Sidney. "Sweetness and Meaning." In Korsmeyer, *The Taste Culture Reader*, 110–22.

Mouristen, Ole G., and Klavs Styrbæk. *Umami: Unlocking the Secrets of the Fifth Taste*. New York: Columbia University Press, 2014.

Moyler, D. A. "Oleoresins, Tinctures, and Extracts." In Ashurst, *Food Flavorings*, 58–84.

Muḥammad ibn al-Ḥasan ibn al-Karīm, *A Baghdad Cookery Book: The Book of Dishes (Kitiib al-Tabfkh)*. Translated by Charles Perry. Blackawton, England: Prospect Books, 2005.

NIIR Board. *Food Flavours Technology Hand Book*. Delhi: National Institute of Industrial Research, 2004.

Norman, Jill. *Herbs and Spices: The Cook's Reference*. New York: Dorling Kindersley, 2002.

O'Neil, Darcy. *Fix the Pumps*. Self-Published, 2009.

Pacult, F. Paul. *Kindred Spirits 2*. Wallkill, NY: Spirit Journal, 2008.

Pasqualini, Dominique T., and Bruno Suet. *The Time of Tea*. Paris: Marval, 1999.

Peterson, T. Sarah. "Food as Divine Medicine." In Korsmeyer, *The Taste Culture Reader*, 147–55.

Peynaud, Émile. "Tasting Problems and Errors of Perception." In Korsmeyer, *The Taste Culture Reader*, 272–84.

Pruthi, J. S. *Spices and Condiments*. New Delhi: National Book Trust India, 1976.

Raghavan, Susheela. *Spices, Seasonings, and Flavorings*. 2nd ed. Boca Raton, FL: CRC Press, 2007.

Redgrove, H. S. *Spices and Condiments*. London: Sir Isaac Pitman & Sons, 1933.

Revel, Jean-François. *Culture and Cuisine*. New York: Doubleday, 1982.

————. "Retrieving Tastes: Two Sources of Cuisine." In Korsmeyer, *The Taste Culture Reader*, 51–56.

Richardson, Tim. *Sweets: A History of Candy*. New York: Bloomsbury, 2003.

Rosenberger, Bernard. "Arab Cuisine and Its Contribution to European Culture." In Flandrin and Montanari, *Food: A Culinary History*, 207–23.

Rozin, Elizabeth. *Ethnic Cuisine: The Flavor-Principle Cookbook*. Brattleboro, VT: Stephen Greene Press, 1983.

Rozin, Elizabeth, and Paul Rozin. "Culinary Themes and Variations." In Korsmeyer, *The Taste Culture Reader*, 34–41.

Sawer, [John Charles]. *Odorographia: A Natural History of Raw Materials and Drugs Used in the Perfume Industry*, 2 vols. London: Gurney & Jackson, 1892, 1894.

Schafer, Edward H. *The Golden Peaches of Samarkand: A Study of T'ang Exotics*. Berkeley: University of California Press, 1963.

Schwartz, Susan L. *Rasa: Performing the Divine in India*. New York: Columbia University Press, 2004.

Segnit, Niki. *The Flavor Thesaurus*. New York: Bloomsbury USA, 2010.

Serres, Michel. *The Five Senses: A Philosophy of Mingled Bodies*. New York: Continuum, 2008.

Smith, Pamela H. *The Body of the Artisan: Art and Experience in the Scientific Revolution*. Chicago: University of Chicago Press, 2004.

Spence, Charles, and Betina Piqueras-Fiszman. *The Perfect Meal: The Multisensory Science of Food and Dining*. Oxford: John Wiley & Sons/Blackwell, 2014.

Stevenson, Richard J. *The Psychology of Flavour*. Oxford: Oxford University Press, 2009.

Sutton, David E. "Synesthesia, Memory, and the Taste of Home." In Korsmeyer, *The Taste Culture Reader*, 304–16.

Symons, Michael. *A History of Cooks and Cooking*. Champaign: University of Illinois Press, 1998.

Tannahill, Reay. *Food in History*. New York: Three Rivers Press, 1989.

Terrington, William. *Cooling Cups and Dainty Drinks*. London: George Routledge, 1869.

BIBLIOGRAPHY

Titley, Norah M. *The Ni'matnama Manuscript of the Sultans of Mandu*. London: Routledge, 2012.

Tuan, Yi-Fu. "Pleasures of the Proximate Senses: Eating, Taste, and Culture." In Korsmeyer, *The Taste Culture Reader*, 226–34.

Visser, Margaret. "Salt: The Edible Rock." In Korsmeyer, *The Taste Culture Reader*, 105–9.

Waley-Cohen, Joanna. "The Quest for Perfect Balance: Taste and Gastronomy in Imperial China." In Freedman, *Food: The History of Taste*, 99–134.

Williams, David G. *The Chemistry of Essential Oils: An Introduction for Aromatherapists, Beauticians, Retailers and Students*. 2nd ed. Weymouth, England: Micelle Press, 2008.

Wright, John. "Essential Oils." In Ashurst, *Food Flavorings*, 25–57.

———. *Flavor Creation*. Carol Stream, IL: Allured Business Media, 2011.

Wulff, Donna M. "Religion in a New Mode: The Convergence of the Aesthetic and the Religious in Medieval India." *Journal of the American Academy of Religion* 54, no. 4 (Winter 1986), 673–88.

LIST OF RECIPES